Contributions to Statistics

For further volumes:
http://www.springer.com/series/2912

Matteo Grigoletto • Francesco Lisi
Sonia Petrone

Editors

Complex Models
and Computational Methods
in Statistics

 Springer

Physica-Verlag
A Springer Company

Editors

Matteo Grigoletto
Francesco Lisi
Department of Statistical Sciences
University of Padua
Padua
Italy

Sonia Petrone
Department of Decision Sciences
Bocconi University
Milan
Italy

In collaboration with Department of Statistical Sciences, University of Padua, Italy

ISSN 1431-1968
ISBN 978-88-470-5565-0   ISBN 978-88-470-2871-5 (eBook)
DOI 10.1007/978-88-470-2871-5
Springer Milan Heidelberg New York Dordrecht London

Jointly published with Physica-Verlag Heidelberg, Germany

Preface

The use of computational methods in statistics to face complex problems and highly dimensional data, as well as the widespread availability of computer technology, is no news. The range of applications, instead, is unprecedented. As often occurs, new and complex data types require new strategies, demanding for the development of novel statistical methods and suggesting stimulating mathematical problems.

This volume presents the revised version of a selection of the papers given at S.Co. 2011, the *7th Conference on Statistical Computation and Complex Systems*, held in Padua, Italy, September 19–21, 2011. The S.Co. conference is a forum for the discussion of new developments and applications of statistical methods and computational techniques for complex and high-dimensional datasets.

Although the topics covered in this volume are diverse, the same themes recur, as research is mostly fueled by the need to analyse complicated data sets, for which traditional methods do not provide viable solutions. Among the topics presented we have estimation of traffic matrices in a communications network, in the presence of long-range dependence; nonparametric mixed-effects models for epidemiology; advanced methods for neuroimaging; efficient computations and inference in environmental studies; hierarchical and nonparametric Bayesian methods with applications in genomic studies; Markov switching models to explain regime changes in the evolution of realized volatility for financial returns; joint modelling of financial returns and multiple daily realized measures; classification of multivariate linear–circular data, with applications to marine monitoring networks; forecasting of electricity supply functions, using principal component analysis and reduced rank regression; clustering based on nonparametric density estimation; surface estimation and spatial smoothing, with applications to the estimation of the blood-flow velocity field. Whilst not exhaustive, this list should give a feel of the range of issues discussed at the conference.

This book is addressed to researchers working at the forefront of the statistical analysis of complex systems and using computationally intensive statistical methods.

We wish to thank all contributors who made this volume possible. Finally, thanks must go to the reviewers, who responded rapidly when put under pressure and helped improve the papers with their valuable comments and suggestions.

Padua, Italy   Matteo Grigoletto, Francesco Lisi
Milan, Italy   Sonia Petrone

Contents

A New Unsupervised Classification Technique Through Nonlinear Non Parametric Mixed-Effects Models

Laura Azzimonti, Francesca Ieva, and Anna Maria Paganoni

Abstract In this work we propose a novel unsupervised classification technique based on the estimation of nonlinear nonparametric mixed-effects models. The proposed method is an iterative algorithm that alternates a nonparametric EM step and a nonlinear Maximum Likelihood step. We apply this new procedure to perform an unsupervised clustering of longitudinal data in two different case studies.

1 Introduction

Unsupervised clustering is one of the main topics in data mining, i.e., the process of finding useful information from data [4]. We focus our attention on highly overdispersed longitudinal and repeated data, which are naturally described through mixed-effects models. Nonlinear mixed-effects models (NLME models) are mixed-effects models in which at least one of the fixed or random effects appears nonlinearly in the model function. They are increasingly used in several biomedical and ecological applications, especially in population pharmacokinetics, pharmacodynamic, immune cells reconstruction and epidemiological studies (see [6, 7, 13, 21]). In these fields, statistical modeling based on NLME models takes advantage of tools that allow to distinguish overall population effects from drugs effects or unit specific influence. In general, mixed-effects models include parameters associated with the entire population (fixed effects) and subject/group specific parameters (random effects). For this reason, mixed-effects models are able to describe the dynamics of the phenomenon under investigation, even in the presence of high between subjects variability. When the random effects represent a deviation from the common dynamics of the population, mixed-effects models

L. Azzimonti · F. Ieva (✉) · A.M. Paganoni
MOX - Department of Mathematics, Politecnico di Milano, Milan, Italy
e-mail: laura.azzimonti@mail.polimi.it; francesca.ieva@mail.polimi.it; anna.paganoni@polimi.it

M. Grigoletto et al. (eds.), *Complex Models and Computational Methods in Statistics*,
Contributions to Statistics, DOI 10.1007/978-88-470-2871-5_1,
© Springer-Verlag Italia 2013

provide estimates both for the entire population's model and for each subject's one. In this work random effects have a different meaning, in fact they describe the common dynamics of different groups of subjects. In this framework, mixed-effects models provide only estimates for each group-specific model. Thanks to this property, it will be possible to consider mixed-effects models as an unsupervised clustering tool for longitudinal data and repeated measures. For this reason we focus our attention on the estimation of the distribution of the random effects $\mathcal{P}^*$.

A wide literature exists for parametric modeling of random effects distribution in linear and non linear mixed-effects models. In this framework, Maximum Likelihood (ML) estimators are generally preferred because of their consistency and efficiency. However, due to the non linearity of the likelihood, we are not always able to provide explicitly the parameter estimators. A general and complete overview of linear multilevel models is given in [12]. An analogous overview for nonlinear case is given in [10]. In [9] it is shown how R and S-plus tools estimate linear and generalized linear mixed-effects models with parametric, in particular Gaussian, random effects. Concerning nonlinear models, in [11] a ML estimation of Gaussian random effect is provided for peculiar nonlinear forms. A stochastic approximation of traditional EM algorithm (SAEM) for estimating Gaussian random effects is suggested in [14], whereas an exact EM algorithm is described in [24]. Finally, [25] introduces a Laplace approximation for nonlinear random effects marginal distributions. However, parametric assumptions may sometimes result too restrictive to describe very heterogeneous or grouped populations. Moreover, when the number of measurements for unit is small, predictions for random effects are strongly influenced by the parametric assumptions. For these reasons nonparametric (NP) framework, which allow $\mathcal{P}^*$ to live in an infinite dimensional space, is attractive. Moreover, it provides in a very natural way a clustering tool, as we will highlight later.

Methods for the estimation of linear nonparametric random effects distribution in linear and generalized linear mixed-effects models have been proposed in [1, 2], whereas [3, 6, 15, 23], among others, deal with nonparametric nonlinear models.

In this work we propose a novel estimation method for nonlinear nonparametric mixed-effects models, aimed at unsupervised clustering. The proposed method is an iterative algorithm that alternates a nonparametric EM step and a nonlinear Maximum Likelihood step. The present algorithm is implemented in R program (version 2.13.0, R Development Core Team [20]) and the R source code is available upon request. To the best of our knowledge, this is the first example of free software for the estimation of nonlinear nonparametric mixed-effects models.

In Sect. 2 the general framework of the work is sketched out, and in Sect. 3 the algorithm for the estimation of nonlinear nonparametric random effect (NLNPEM) is described. Section 4 contains applications to case studies. Concluding remarks and further developments of this work are finally discussed in Sect. 5. Technical details in the estimation algorithm are discussed in Appendix.

2 Model and Framework

We consider the following NLME model for longitudinal data:

$$\mathbf{y}_i = f(\boldsymbol{\beta}, \mathbf{b}_i, \mathbf{t}) + \boldsymbol{\epsilon}_i \quad i = 1, \ldots, N$$
$$\boldsymbol{\epsilon}_i \sim \mathcal{N}(\mathbf{0}, \sigma^2 \mathbb{I}_n) \quad \text{i.i.d.} \tag{1}$$

where $\mathbf{y}_i \in \mathbb{R}^n$ is the response variable evaluated at times $\mathbf{t} \in \mathbb{R}^n$ and f is a general, real-valued and differentiable function with $p + q$ parameters. Each parameter of f is treated either as fixed or as random. Fixed effects are parameters associated with the entire population whereas random effects are subject-specific parameters that allow to identify clusters of subjects. $\boldsymbol{\beta} \in \mathbb{R}^p$ is a vector that contain all fixed effects and $\mathbf{b}_i \in \mathbb{R}^q$ is the vector for the i-th subject random effects.

The function f is nonlinear at least in one component of the fixed or random effects. The errors ϵ_{ij} are associated with the j-th measurement of the i-th longitudinal data. They are normally distributed, independent between different subjects and independent within the same subject. In general, the proposed method could also take into account of a different number of observations, located at different times, for different subjects. In (1) we chose not to consider this case in order to ease the notation, but the generalization is straightforward.

Usually random effects are assumed to be Normal distributed, $\mathbf{b}_i \sim \mathcal{N}_q(\mathbf{0}, \Sigma)$, with unknown parameters that, together with $\boldsymbol{\beta}$ and σ, can be estimated through methods based on the likelihood function (see [18]). In this parametric framework the maximum likelihood estimators are generally favored by their statistical properties, i.e., consistency and efficiency. Nevertheless the parametric assumptions could be too restrictive to describe highly heterogeneous or grouped data, so it might be necessary to move to a nonparametric approach. In our case, we assume $\mathbf{b}_i$, for $i = 1, \ldots, N$, independent and identically distributed according to a probability measure $\mathcal{P}^*$ that belongs to the class of all probability measures on $\mathbb{R}^q$. $\mathcal{P}^*$ can then be interpreted as the mixing distribution that generates the density of the stochastic model in (1). Looking for the ML estimator $\hat{\mathcal{P}}^*$ of $\mathcal{P}^*$ in the space of all probability measures on $\mathbb{R}^q$, the discreteness theorem proved in [16] states that $\hat{\mathcal{P}}^*$ is a discrete measure with at most N support points. Moreover under suitable hypotheses on the distribution of the response variable, satisfied, for example, by densities in the exponential family, the ML estimator is also unique as proved in [17]. Therefore the ML estimator of the random effects distribution can be expressed as a set of points $(\mathbf{c}_1, \ldots, \mathbf{c}_M)$, where $M \leq N$ and $\mathbf{c}_l \in \mathbb{R}^q$, and a set of weights $(\omega_1, \ldots, \omega_M)$, where $\omega_l \geq 0$ and $\sum_{l=1}^{M} \omega_l = 1$.

As mentioned above, in this paper we propose an algorithm for the joint estimation of $\boldsymbol{\beta}$, $(\mathbf{c}_1, \ldots, \mathbf{c}_M)$, $(\omega_1, \ldots, \omega_M)$ and σ^2 in the nonlinear framework of model (1). The proposed method maximizes the following likelihood

$$L(\boldsymbol{\beta}, \sigma^2 | \mathbf{y}) = p(\mathbf{y} | \boldsymbol{\beta}, \sigma^2) = \sum_{l=1}^{M} \omega_l \frac{1}{(2\pi\sigma^2)^{(nN)/2}} e^{-\frac{1}{2\sigma^2} \sum_{i=1}^{N} \sum_{j=1}^{n} \left(y_{ij} - f(\boldsymbol{\beta}, \mathbf{c}_l, t_j)\right)^2}$$

$$\tag{2}$$

with respect to fixed effects $\boldsymbol{\beta}$, error variance σ^2, and the random effects distribution $(\mathbf{c}_l, \omega_l)$, $l = 1, \ldots, M$. Each iteration of the algorithm described in Sect. 3 increases the likelihood in (2).

Concerning the distribution of random effects, for each $l = 1, \ldots, M$, $\mathbf{c}_l$ and ω_l represent the group-specific parameters and the corresponding weights in the mixture (2), respectively. Notice that we do not have to fix a priori the number M of support points, but it is computed by the algorithm. Since we don't have to specify a priori the number of support points and in consequence the number of groups, the nonparametric mixed-effects model could be interpreted as an unsupervised clustering tool for longitudinal data. This tool could be very useful in order to identify groups of subjects to be used in the analysis and to cluster observations.

3   NLNPEM Algorithm

The algorithm proposed for the estimation of the parameters of model (1) arises from the framework described in [22], and it increases at each iteration the likelihood (2). The algorithm alternates two steps: the first one is a nonparametric EM step whereas the second one is a nonlinear maximum-likelihood step. The nonparametric EM step estimates the discrete q-dimensional distribution $(\mathbf{c}_l, \omega_l)$, $l = 1, \ldots, M$ of the random effects $\mathbf{b}_i$. The non linear maximum likelihood step provides an estimation of the fixed effects $\boldsymbol{\beta}$ and the variance σ^2, given $\mathbf{b}_i$.

The nonparametric EM step consists in an update of the parameters of the discrete distribution $(\mathbf{c}_l, \omega_l)$, $l = 1, \ldots, M$ that increases the likelihood function (2). The property of increasing the likelihood was proved in [22]. The update is the following:

$$
\begin{cases}
\omega_l^{\text{up}} = \dfrac{1}{N} \sum_{i=1}^{N} W_{il} \\[2ex]
\mathbf{c}_l^{\text{up}} = \arg\max_{\mathbf{c}} \left[\sum_{i=1}^{N} W_{il} \ln p(\mathbf{y}_i \mid \boldsymbol{\beta}, \sigma^2, \mathbf{c}) \right]
\end{cases}
\tag{3}
$$

where

$$
W_{il} = \frac{\omega_l \, p(\mathbf{y}_i \mid \boldsymbol{\beta}, \sigma^2, \mathbf{c}_l)}{\sum_{k=1}^{M} \omega_k \, p(\mathbf{y}_i \mid \boldsymbol{\beta}, \sigma^2, \mathbf{c}_k)}
$$

and

$$
p(\mathbf{y}_i \mid \boldsymbol{\beta}, \sigma^2, \mathbf{c}) = \frac{1}{(2\pi\sigma^2)^{n/2}} e^{-\frac{1}{2\sigma^2} \sum_{j=1}^{n} \left(y_{ij} - f(\boldsymbol{\beta}, \mathbf{c}, t_j) \right)^2}.
$$

The coefficients W_{il} represent the probability of $\mathbf{b}_i$ being equal to $\mathbf{c}_l$ conditionally to the observation $\mathbf{y}_i$ and given the fixed effects $\boldsymbol{\beta}$ and the variance σ^2, that is

$$
W_{il} = p(\mathbf{b}_i = \mathbf{c}_l \mid \mathbf{y}_i, \boldsymbol{\beta}, \sigma^2)
$$

in fact,

$$W_{il} = \frac{p(\mathbf{b}_i = \mathbf{c}_l)p(\mathbf{y}_i \mid \boldsymbol{\beta}, \sigma^2, \mathbf{c}_l)}{p(\mathbf{y}_i \mid \boldsymbol{\beta}, \sigma^2)} = \frac{p(\mathbf{y}_i, \mathbf{b}_i = \mathbf{c}_l \mid \boldsymbol{\beta}, \sigma^2)}{p(\mathbf{y}_i \mid \boldsymbol{\beta}, \sigma^2)} = p(\mathbf{b}_i = \mathbf{c}_l \mid \mathbf{y}_i, \boldsymbol{\beta}, \sigma^2).$$

In order to estimate $\mathbf{b}_i$ for $i = 1, \dots, N$, we want to maximize the probability of $\mathbf{b}_i$ conditionally to the observations $\mathbf{y}_i$ and given the fixed effects $\boldsymbol{\beta}$ and the error variance σ^2. For this reason the estimation of the random effects, $\hat{\mathbf{b}}_i$, is obtained maximizing W_{il} over l, that is

$$\hat{\mathbf{b}}_i = \mathbf{c}_{\bar{l}} \text{ if } \bar{l} = \arg\max_l W_{il}.$$

During the nonparametric EM step, we could also reduce the support of the discrete distribution. The reduction of the support is performed in order to cluster the random effects. This support reduction consists in both making points very close to each other collapse and removing points with very low weight and not associated with any subject. In particular if two points are too close, that is $\|\mathbf{c}_l - \mathbf{c}_k\| < D$, where D is a tuning tolerance parameter, then we replace $\mathbf{c}_l$ and $\mathbf{c}_k$ with a new point $\mathbf{c}_{\min\{l,k\}} = (\mathbf{c}_l + \mathbf{c}_k)/2$ with weight $\omega_{\min\{l,k\}} = \omega_l + \omega_k$. Otherwise, if $\omega_l < \tilde{\omega}$, where $\tilde{\omega}$ is another tuning tolerance parameter, and the subset $\{i : \hat{\mathbf{b}}_i = \mathbf{c}_l\}$ is empty, we remove the point $\mathbf{c}_l$. The thresholds D and $\tilde{\omega}$ are two complexity parameters that affect the estimation of the nonparametric distribution; $\tilde{\omega}$ is linked to the size of the smallest group that we want to detect, while D represents the minimum allowed distance between different points of the discrete random effects distribution; the higher D is set, the lower is the number of groups. For this reason the two complexity parameters define a trade-off between bias and high number of groups. In this work we prefer setting D low in order to obtain a higher number of groups and, in case, cluster them later. A rule of thumb for setting these threshold parameters is the following: D may be much smaller than the standard deviation within groups, on the other hand, $\tilde{\omega}$ may be set of the same order of the inverse of the total number of observations in the dataset.

The nonlinear maximum likelihood step provides the estimation of the fixed effects $\boldsymbol{\beta}$ and the errors variance σ^2, given $\mathbf{b}_i = \hat{\mathbf{b}}_i$. In this step we maximize the nonlinear log-likelihood:

$$\ell(\boldsymbol{\beta}, \sigma^2 \mid \mathbf{y}, \hat{\mathbf{b}}) = -\frac{nN}{2}\ln(2\pi\sigma^2) - \frac{1}{2\sigma^2}\sum_{i=1}^{N}\sum_{j=1}^{n}\left(y_{ij} - f(\boldsymbol{\beta}, \hat{\mathbf{b}}_i, t_j)\right)^2 \qquad (4)$$

where $\hat{\mathbf{b}}_i$ is the estimation of random effects for the i-th subject provided in the nonparametric EM step.

The algorithm, given a starting discrete distribution with N support points for the random effects and a starting estimate for the fixed effects, alternates the nonparametric EM step and the nonlinear maximum likelihood step until convergence. More details, together with the sketch of the algorithm, are reported in Appendix.

In order to validate the proposed estimation algorithm and to compare it with already existing procedures for the linear framework, an intensive simulation study has been performed and detailed in [5]. In the first simulation study (see [5], Sect. 3.2), we compared the results obtained in a linear framework with those obtained with the algorithm introduced in [1] and implemented in the npmlreg R-package (see [8]). In the second one (see [5], Sect. 3.3) we considered two classic nonlinear functions f in (1): the exponential and the logistic growth curves. For each case a test set of simulated curves has been designed and the algorithm performance in the estimation of the random effects has been evaluated computing the Wasserstein distance between the true and the estimated distribution of the random effects.

In the linear framework NLNPEM method performs very well and its results are comparable with those obtained with the already existing npmlreg method; for a large number of groups, npmlreg method doesn't detect some points of the nonparametric distribution or even doesn't reach convergence, whereas NLNPEM performs well, even ignoring the true number of groups. Both in the linear and nonlinear framework we obtain a very high level of agreement, measured in term of Wasserstein distance between the true distribution generating data and the one estimated by NLNPEM algorithm. The NLNPEM method is also able to capture correctly outlier groups even in highly unbalanced situations.

4 Application to Case Studies

In this section we apply the proposed method to two different datasets: the first one contains the carbon dioxide uptake photosynthetic response curves in a sample of 12 different plants. It is a classical dataset for the study of longitudinal curves presented in [19] in a study of the cold tolerance of a C_4 grass species, *Echinochloa crusgalli* and analyzed also in [18]. The second one describes the number of Hospital Discharges of patients affected by Acute Myocardial Infarction (AMI) without ST-segment Elevation (NON-STEMI) along the time period 2000–2007, grouped by hospital and relative to the 30 largest clinical institutions of Regione Lombardia. The explorative analysis of these data is aimed at detecting groups with similar behaviours.

4.1 Carbon Dioxide Uptake

In the first case we consider the carbon dioxide (CO_2) uptake [$\mu mol \cdot m^{-2} \cdot s^{-1}$] of 12 plants, measured at several levels of ambient CO_2 concentration [$\mu L/L$], see Fig. 1. In [19] an exponential growth model is proposed to capture the common shape of the curves. In this case the nonlinear function to be used in the model (1) is:

$$f(t) = \alpha \left(1 - e^{-\lambda t}\right)$$

Fig. 1 Carbon dioxide uptake photosynthetic curves for 12 plants. Real data are colored according to the NLNPEM clusters and NLNPEM fitted models are superimposed

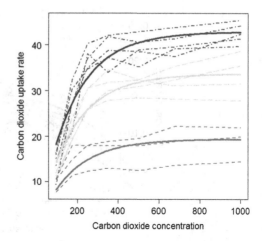

which is nonlinear in λ. The two parameters α and λ represent, respectively, the asymptote and the growth rate.

In this analysis we consider only random effects for the asymptote, that means that the mixed-effects model becomes

$$\mathbf{y}_i = a_i \left(1 - e^{-\lambda t}\right) + \boldsymbol{\epsilon}_i$$

where $\boldsymbol{\epsilon}_i \sim \mathcal{N}(\mathbf{0}, \sigma^2 \mathbb{I}_n)$ are i.i.d. errors, a_i are the random effects for the asymptote ($b_i = a_i$), and λ is the fixed effect for the growth rate ($\beta = \lambda$).

The NLNPEM algorithm clusters the plants in $M = 3$ different groups, according to the estimated discrete distribution of the random effect for the asymptote (see Fig. 1). The estimated fixed effect is $\hat{\lambda} = 0.006$, the estimated discrete measure $\hat{\mathcal{P}}^*$ is concentrated on $(\hat{\mathbf{c}}_1, \hat{\mathbf{c}}_2, \hat{\mathbf{c}}_3) = (19.39, 33.71, 42.89)$ with weights $(\hat{\omega}_1, \hat{\omega}_2, \hat{\omega}_3) = (0.25, 0.33, 0.42)$ and the estimated variance is $\hat{\sigma}^2 = 8.94$. This analysis, performed with $D = 5$ and $\tilde{\omega} = 0.05$, backs up the presence of three groups of plants according to different asymptotes and automatically detects an unsupervised cluster structure. This result is in total agreement with a k-means clustering of the random asymptote point estimates computed following the traditional parametric approach [18] that assumes a Normal model for the random effect. Nevertheless, in that case, a critical point is the choice of k, the number of groups, which is set equal to three after maximizing the average silhouette width. On the contrary the number of groups is automatically computed in the NLNPEM method.

4.2 *Acute Myocardial Infarction Without ST-Segment Elevation*

The second example analyzed comes from epidemiological studies carried out using administrative databanks. In fact, Fig. 2 represents the normalized number of NON-STEMI diagnoses along the time period 2000–2007 grouped by hospital

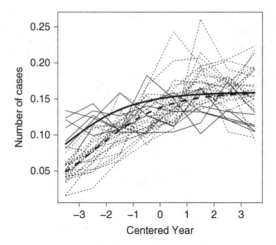

Fig. 2 Standardized number of AMI without ST-segment elevation diagnoses in the period 2000–2007 in the 30 largest clinical institutions of Lombardia Region. The year has been centered and normalization has been carried out standardizing the yearly number of diagnoses for each hospital by total number of diagnoses in the time window 2000–2007. Clusters pointed out by NLNPEM algorithm are highlighted, respectively, by *solid* and *dashed lines*. NLNPEM fitted models are superimposed

and relative to the 30 largest clinical institutions of Regione Lombardia. For each hospital the yearly number of diagnoses has been standardized by the hospital total number of diagnoses in the time period 2000–2007. As pointed out in [13] a logistic growth model with random inflection seems to capture the common "S-shaped" growing pattern; for this reason we consider a logistic growth model. In this case, the nonlinear function to be used in model (1) is:

$$f(t) = \frac{\alpha}{1 + e^{-\frac{t-\delta}{\gamma}}}$$

where α represent the asymptote, δ is the inflection point, which correspond to the time at which the growth curve reaches the half of the asymptote, and γ is the time elapsed between δ and the time at which the growth curve reaches 3/4 of the asymptotic level. The parameters γ and α are treated as fixed effects while the inflection point is treated as random, as suggested in [13]. The model becomes:

$$\mathbf{y}_i = \frac{\alpha}{1 + e^{-\frac{t-d_i}{\gamma}}} + \boldsymbol{\epsilon}_i$$

where $\boldsymbol{\epsilon}_i \sim \mathcal{N}(\mathbf{0}, \sigma^2 \mathbb{I}_n)$ are i.i.d. errors, d_i represent the random effects for the inflection point, while α and γ represent the fixed effects. In particular $b_i = d_i$ and $\boldsymbol{\beta} = (\alpha, \gamma)$.

The NLNPEM algorithm clusters the hospitals in $M = 2$ different groups, according to the estimated discrete distribution of the random inflection point (see Fig. 2). The estimated fixed effects are $\hat{\alpha} = 0.16$ and $\hat{\gamma} = 1.31$, the estimated discrete measure $\hat{\mathcal{P}}^*$ is concentrated on $(\hat{\mathbf{c}}_1, \hat{\mathbf{c}}_2) = (-3.76, -2.43)$ with weights $(\hat{\omega}_1, \hat{\omega}_2) = (0.2, 0.8)$ and the estimated variance is $\hat{\sigma}^2 = 7.7 \cdot 10^{-4}$. This analysis, performed with $D = 0.05$ and $\tilde{\omega} = 0.05$, backs up the presence of two groups of hospitals according to different inflection points.

Even if clinical best practice maintains that there is no evidence for a greater incidence of NON-STEMI in this period, it is known that since the early 2000s a new diagnostic procedure—the *troponin* exam—has been introduced and this could have produced an increased number of positive diagnoses by easing NON-STEMI detection. Hence, the presence of two clusters could be a consequence of the different hospital timings in the introduction and adoption of this practice. This hypothesis cannot be validated directly since the timings of adoption of the troponin exam by the 30 different hospitals included in the analysis are not available.

The good agreement with previous results detailed in [13] together with the great advantage of a nonparametric approach advocates the real profit in using this new estimation algorithm.

5   Conclusions

In this work, we proposed a new unsupervised clustering technique based on a new estimation method for nonlinear nonparametric mixed-effects models. The proposed method is based on an iterative algorithm (named NLNPEM) that alternates a nonparametric EM and an optimization step for the maximization of a nonlinear likelihood function. A simulation study both in linear and nonlinear setting of exponential and logistic growth has been carried out. Results show that NLNPEM performs well, even ignoring the real number of groups, in terms of Wasserstein distance between the true distribution generating data and the one estimated by NLNPEM algorithm, and that it always reaches convergence, even in those cases where several groups are present. We use this algorithm as an unsupervised clustering technique in the context of the explorative data mining. In particular two applications to real data of carbon dioxide uptake photosynthetic response curves and NON-STEMI number of diagnoses, respectively, are presented. In these two case studies the potential of our method in unsupervised clustering analysis is highlighted.

Appendix: Details on NLNPEM Algorithm

The NLNPEM is the following:

1. Define a starting discrete distribution for random effects with support on N points $(\mathbf{c}^{(0)}, \omega^{(0)})$, a starting estimate for the fixed effects $\boldsymbol{\beta}^{(0)}$ and for $\sigma^{2(0)}$ and the tolerance parameters D and $\tilde{\omega}$.

2. Given $(\mathbf{c}^{(k-1)}, \boldsymbol{\omega}^{(k-1)})$, $\boldsymbol{\beta}^{(k-1)}$ and $\sigma^{2(k-1)}$, perform the EM step (without the support reduction) in order to update the support points $\mathbf{c}^{(k)}$ and the weights $\boldsymbol{\omega}^{(k)}$ of the random effect distribution, according to (3).
3. Given $(\mathbf{c}^{(k)}, \boldsymbol{\omega}^{(k)})$, perform the nonlinear maximum likelihood step in order to estimate the fixed effects $\boldsymbol{\beta}^{(k)}$ and the error variance $\sigma^{2(k)}$ maximizing (4).
4. Iterate Steps 2 and 3 until convergence.
5. Reduce the support of the discrete distribution, according to the tuning parameters D and $\tilde{\omega}$.
6. Given $(\mathbf{c}^{(k-1)}, \boldsymbol{\omega}^{(k-1)})$, $\boldsymbol{\beta}^{(k-1)}$, $\sigma^{2(k-1)}$, D and $\tilde{\omega}$, perform the EM step with the support reduction in order to update the support points $\mathbf{c}^{(k)}$ and the weights $\boldsymbol{\omega}^{(k)}$ of the random effect distribution, according to (3).
7. Given $(\mathbf{c}^{(k)}, \boldsymbol{\omega}^{(k)})$, perform the nonlinear maximum likelihood step in order to estimate the fixed effects $\boldsymbol{\beta}^{(k)}$ and the error variance $\sigma^{2(k)}$ maximizing (4).
8. Iterate Steps 6 and 7 until convergence.

The algorithm reaches convergence when parameters and discrete distribution stop changing or when there is no variation in the log-likelihood function.

Acknowledgments The case study in Sect. 4 is within the Strategic Program "Exploitation, integration and study of current and future health databases in Lombardia for Acute Myocardial Infarction" supported by "Ministero del Lavoro, della Salute e delle Politiche Sociali" and by "Direzione Generale Sanità—Regione Lombardia."

References

1. Aitkin, M.: A general maximum likelihood analysis of overdispersion in generalized linear models. Stat. Comput. **6**, 251–262 (1996)
2. Aitkin, M.: A general maximum likelihood analysis of variance components in generalized linear models. Biometrics **55**, 117–128 (1999)
3. Antic, J., Laffont, C.M., Chafaï, D., Concordet, D.: Comparison of nonparametric methods in nonlinear mixed effect models. Comput. Stat. Data Anal. **53**(3), 642–656 (2009)
4. Azzalini, A., Scarpa, B.: Data Analysis and Data Mining. Oxford Univeristy Press, Oxford (2012)
5. Azzimonti, L., Ieva F., Paganoni, A.M.: Nonlinear nonparametric mixed-effects models for unsupervised classification. Computational Statistics, Published online: 27 September 2012. DOI: 10.1007/s00180-012-0366-5
6. Davidian, M., Gallant, A.R.: The nonlinear mixed effects model with a smooth random effects density. Biometrika **80**(3), 475–488 (1993)
7. De Lalla, C., Rinaldi, A., Montagna, D., Azzimonti, L., Bernardo, M.E., Sangalli, L.M., Paganoni, A.M., Maccario, R., Di Cesare Merlone, A., Zecca, M., Locatelli, F., Dellabona, P., Casorati, G.: Invariant natural killer T-cell reconstitution in pediatric leukemia patients given HLA-haploidentical stem cell transplantation defines distinct CD4+ and CD4− subset dynamics and correlates with remission state. J. Immunol. **186**(7), 4490–4499 (2011)
8. Einbeck, J., Darnell, R., Hinde, J.: npmlreg: nonparametric maximum likelihood estimation for random effect models. [Online] http://CRAN.R-project.org/package=npmlreg (2009) (Accessed: 26 November 2012)

9. Fox, J.: Linear mixed models. Appendix to An R and S-PLUS Companion to Applied Regression. Sage Publications Inc. California (2002) http://cran.r-project.org/doc/contrib/Fox-Companion/appendix.html (Accessed: 26 November 2012)
10. Gallant, A.R.: Nonlinear Statistical Models. Wiley, New York (1987)
11. Goldstein, H.: Nonlinear multilevel models, with an application to discrete response data. Biometrika **78**(1), 45–51 (1991)
12. Hox, J.J.: Applied Multilevel Analysis. TT-Publikaties, Amsterdam (1995)
13. Ieva, F., Paganoni, A.M., Secchi, P.: Mining administrative health databases for epidemiological purposes: a case study on Acute Myocardial Infarctions diagnoses. In: Pesarin, F., Torelli, S. (eds.) Accepted for Publication in Advances in Theoretical and Applied Statistics. Springer, Berlin (2012) http://mox.polimi.it/it/progetti/pubblicazioni/quaderni/45-2010.pdf
14. Kuhn, E., Lavielle, M.: Maximum likelihood estimation in nonlinear mixed effect models. Comput. Stat. Data Anal. **49**(4), 1020–1038 (2005)
15. Lai, T.L., Shih, M.C.: Nonparametric estimation in nonlinear mixed-effects models. Biometrika **90**(1), 1–13 (2003)
16. Lindsay, B.G.: The geometry of mixture likelihoods: a general theory. Ann. Stat. **11**(1), 86–94 (1983a)
17. Lindsay, B.G.: The geometry of mixture likelihoods, Part II: the exponential family. Ann. Stat. **11**(3), 783–792 (1983b)
18. Pinheiro, J.C., Bates, D.M.: Mixed-Effects Models in S and S-Plus. Springer, Berlin (2000)
19. Potvinm, C., Lechowicz, M.J., Tardif, S.: The statistical analysis of ecophysiological response curves obtained form experiments involving repeated measures. Ecology **71**(4), 1389–1400 (1990)
20. R Development Core Team: R: a language and environment for statistical computing. R Foundation for Statistical Computing, Vienna, Austria [Online] http://www.R-project.org (2009) (Accessed: 26 November 2012)
21. Sheiner, L.B., Beal, S.L.: Evaluation of methods for estimating population pharmacokinetic parameters. III. Monoexponential model: routine clinical pharmacokinetic data. J. Pharmacokinet. Pharmacodyn. **11**(3), 303–319 (1980)
22. Schumitzky, A.: Nonparametric EM algorithms for estimating prior distributions. Appl. Math. Comput. **45**(2), 143–157 (1991)
23. Vermunt, J.K.: An EM algorithm for the estimation of parametric and nonparametric hierarchical models. Statistica Neerlandica **58**, 220–233 (2004)
24. Walker, S.: An EM algorithm for nonlinear random effects models. Biometrics **52**(3), 934–944 (1996)
25. Wolfinger, R.: Laplace's approximation for nonlinear mixed models. Biometrika **80**, 791–795 (1993)

Estimation Approaches for the Apparent Diffusion Coefficient in Rice-Distributed MR Signals

Stefano Baraldo, Francesca Ieva, Luca Mainardi, and Anna Maria Paganoni

Abstract The Apparent Diffusion Coefficient (ADC) is often considered in the differential diagnosis of tumors, since the analysis of a field of ADCs on a particular region of the body allows to identify regional necrosis. This quantity can be estimated from magnitude signals obtained in diffusion Magnetic Resonance (MR), but in some situations, like total body MRs, it is possible to repeat only few measurements on the same patient, thus providing a limited amount of data for the estimation of ADCs. In this work we consider a Rician distributed magnitude signal with an exponential dependence on the so-called b-value. Different pixelwise estimators for the ADC, both frequentist and Bayesian, are proposed and compared by a simulation study, focusing on issues caused by low signal-to-noise ratios and small sample sizes.

1 Introduction

Diffusion magnetic resonance (MR) is as an important tool in clinical research, as it allows to characterize some properties of biological tissues. When tumor areas are analyzed using this technique, it can be observed that the diffusion tensor, estimated from the magnetic MR magnitude signal, has reduced values in lesions with respect to surrounding physiological tissues, allowing to identify pathological areas or necrosis. When the tissue region of interest can be considered as isotropic the *Apparent Diffusion Coefficient* (ADC) is sufficient to characterize the diffusion

S. Baraldo (✉) · F. Ieva · A.M. Paganoni
MOX - Department of Mathematics, Politecnico di Milano, Milan, Italy
e-mail: stefano1.baraldo@mail.polimi.it; francesca.ieva@mail.polimi.it;
anna.paganoni@polimi.it

L. Mainardi
Department of Bioengineering, Politecnico di Milano, Milan, Italy
e-mail: luca.mainardi@polimi.it

M. Grigoletto et al. (eds.), *Complex Models and Computational Methods in Statistics*,
Contributions to Statistics, DOI 10.1007/978-88-470-2871-5_2,
© Springer-Verlag Italia 2013

properties of the tissue, and it is usually estimated from the exponential decay of the signal with respect to the *b-value*, the MR acquisition parameter. The assumption of isotropy is common and reasonable in various cases, like breast and prostate cancer (see, for example, [7, 10]).

In many practical situations it may not be possible to collect more than few measures at different b-values, limiting the accuracy of the estimation. A reduction in the total number of measures necessary to achieve a certain accuracy is convenient in term of costs and allows to keep the patient involved in the MR procedure for a shorter amount of time (the experience may be unpleasant, especially when total body MR must be performed). The purpose of this work is to compare different frequentist and Bayesian approaches to the estimation of the ADC, underlining their statistical properties and computational issues.

2 Rice-Distributed Diffusion MR Signals

2.1 The Rice Distribution

The random variables we deal with derive from the complex signal $w = w_r + iw_i$ measured in diffusion MR. It is usual to assume that both w_r and w_i are affected by a Gaussian noise with equal, constant variance, i.e. $w_r \sim \mathcal{N}(v \cos(\vartheta), \sigma^2)$ and $w_i \sim \mathcal{N}(v \sin(\vartheta), \sigma^2)$, with $v \in \mathbb{R}^+$ and $\vartheta \in [0, 2\pi)$. The quantity at hand is the modulus M of this signal, which has then a *Rice* (or *Rician*) distribution, that we will denote as $M \sim \text{Rice}(v, \sigma^2)$. The density of this random variable has the form

$$f_M(m|v, \sigma^2) = \frac{m}{\sigma^2} e^{-\frac{m^2+v^2}{2\sigma^2}} I_0 \left(\frac{mv}{\sigma^2} \right) \mathbb{I}_{(0,+\infty)}(m), \tag{1}$$

where I_0 is the zeroth-order modified Bessel function of the first kind (see [1]). Using the series expression of I_0, it is possible to deduce a different, equivalent definition of a Rician random variable as $M = \sigma \sqrt{R}$, where R is a noncentral χ^2 variable that can be expressed as a mixture of $\chi^2(2P + 2)$ distributions with $P \sim Poisson(v^2/2\sigma^2)$. This formulation becomes particularly useful for sampling from a Rice distribution, as it allows an easy implementation of a Gibbs sampler.

2.2 Rice Exponential Regression

Diffusion MR aims at computing the diffusion tensor field on a portion of tissue, and this is achieved by analyzing the influence of water diffusion on the measured signal, under different experimental settings. In particular, the classical model for relating the magnitude signal to the acquisition parameters and the 3-dimensional diffusion tensor D is the *Stejskal–Tanner* equation

$$\nu_{\mathbf{g}} = \nu_0 \exp(-\mathbf{g}^{\mathrm{T}} D \mathbf{g} b), \qquad (2)$$

where $\nu_{\mathbf{g}}$ is the "real" intensity signal we want to measure, ν_0 is the signal at $b = 0$ and the vector $\mathbf{g} \in \mathbb{R}^3$ is the applied magnetic gradient. The b-value is a function of other acquisition settings, which we will omit since their description and discussion is beyond the scope of this article. See, for example, [3] for an overview on MR techniques, including diffusion MR, and a discussion of various issues and recent advances in this field.

In general, even in the ideal noiseless case, at least six observations are needed to determine the components of the symmetric, positive definite diffusion tensor D, by varying the direction $\mathbf{g}$ of the magnetic field gradient. However, if the tissue under study can be considered as isotropic, the diffusion tensor has the simpler form $D = \alpha I$, where α is the ADC, a scalar parameter, and I is the identity matrix. This reduces model (2) to the following

$$\nu = \nu_0 \exp(-\alpha b) \qquad (3)$$

for any vector $\mathbf{g}$ (in the following, we will omit it for ease of notation).

Equation (3) describes pointwise the phenomenon on the tissue region of interest. In this study we consider the pixels of a diffusion MR sequence of images as independent and focus on the estimation problem for a single point in space. We do not consider a spatial modeling for the ADC field: although it could be a useful way to filter noise and to capture underlying tissue structures; on the other hand, for diagnostic purposes it may be preferable to submit to the physician an estimate that has not been artificially smoothed.

3 Estimation Methods

In this section we present different methods for the estimation of α, the unknown parameter of interest. We consider a sample of signal intensities on a single pixel $M_i \sim \mathrm{Rice}(\nu_0 e^{-\alpha b_i}, \sigma^2)$, $i = 1, \ldots, n$, and their respective realizations $\mathbf{m} = m_1, \ldots, m_n$ at b-values $\mathbf{b} = b_1, \ldots, b_n$.

The dispersion parameter σ^2 is usually measured over regions where almost pure noise is observed, and used as a known parameter in the subsequent estimates. This estimate of σ^2 is considered as reliable, since it can be based on a very large number of pixels, so we will consider the case of known dispersion parameter.

We consider nonlinear least squares, maximum likelihood and three Bayesian point estimators. In the case of a simple $\mathrm{Rice}(\nu, \sigma^2)$ random variable an iterative method of moments estimator has been proposed in [2], but this technique has no straightforward extension to the case of covariate-dependent ν, while moment equations would be difficult to invert in the considered case. Moreover, under the model assumptions presented in Sect. 2 a decoupling of noise and signal in the fashion of signal detection theory could not be pursued (see [4] for a brief presentation and a Bayesian implementation of the SDT classical scheme).

The different estimation methods for the couple (v_0, α) presented here will be tested under different signal-to-noise ratios (SNRs) v/σ in Sect. 4.

3.1 Nonlinear Regression

A standard approach for the estimation of v_0 and α is to solve a nonlinear least squares problem, which is equivalent to approximating $M_i = v_0 \exp(-\alpha b_i) + \varepsilon_i$ for $i = 1, \ldots, n$, where ε_i are iid, zero mean, Gaussian noise terms. The estimators $\hat{v}_0^{\mathrm{LS}}$ and $\hat{\alpha}^{\mathrm{LS}}$ are defined as

$$(\hat{v}_0^{\mathrm{LS}}, \hat{\alpha}^{\mathrm{LS}}) = \operatorname*{argmin}_{(v_0, \alpha)} \sum_{i=1}^{n} (m_i - v_0 e^{-\alpha b_i})^2,$$

for $v_0, \alpha > 0$, which is equivalent to the solution of the following equations

$$\begin{cases} v_0 \sum_{i=1}^{n} e^{-2\alpha b_i} = \sum_{i=1}^{n} m_i e^{-\alpha b_i}, \\ v_0 \sum_{i=1}^{n} b_i e^{-2\alpha b_i} = \sum_{i=1}^{n} m_i b_i e^{-\alpha b_i}. \end{cases} \tag{4}$$

The approximation to a nonlinear regression model is inconsistent with the phenomenon under study, most evidently for the fact that in this case the noise term is symmetric and it can assume real values. This inconsistency is negligible for high SNR values, since a $\mathrm{Rice}(v, \sigma^2)$ distribution in this case approaches a $\mathcal{N}(v_0, \sigma^2)$, but becomes important with "intermediate" and low SNRs. In [9], the behavior of the Rice distribution with fixing $\sigma = 1$ and varying v is examined, observing that normality can be considered a good approximation at about $v/\sigma > 2.64$, but the sample variance approaches σ^2 only for SNR values greater than 5.19. Even for pixels with high SNRs at $b = 0$, for large b-values the real signal could reach the same order of magnitude of noise, depending on the unknown value of α, and this could lead to very biased estimates. However, the least squares approach is computationally simpler and quicker to carry out, since it can be seen from (4) that v_0 can be expressed as a function of α, thus requiring just a one-dimensional optimization to compute the estimates.

3.2 Maximum Likelihood

The maximum likelihood approach allows to take into account the asymmetry of the signal distribution, always providing admissible values of the parameters. The objective function is the log-likelihood

$$l(v_0, \alpha | \mathbf{m}, \mathbf{b}, \sigma^2) = \log L(v_0, \alpha | \mathbf{m}, \mathbf{b}, \sigma^2) = \sum_{i=1}^{n} \log f_{M_i}(m_i | v_0 e^{-\alpha b_i}, \sigma^2)$$

$$\propto -\frac{1}{2\sigma^2} \sum_{i=1}^{n} v_0^2 e^{-2\alpha b_i} + \sum_{i=1}^{n} \log \left[I_0 \left(\frac{m_i v_0 e^{-\alpha b_i}}{\sigma^2} \right) \right],$$

where f_{M_i} is the Rice density (1), for $i = 1, \ldots, n$. The ML estimator is then

$$(\hat{v}_0^{ML}, \hat{\alpha}^{ML}) = \underset{(v_0, \alpha)}{\operatorname{argmax}} \, l(v_0, \alpha | \mathbf{m}, \mathbf{b}, \sigma^2),$$

for $v_0, \alpha > 0$.

Looking for stationary points of l and using the fact that $I_0'(x) = I_1(x)$, we obtain the following estimating equations

$$\begin{cases} v_0 \sum_{i=1}^{n} e^{-2\alpha b_i} = \sum_{i=1}^{n} \frac{I_1(\frac{m_i v_0 e^{-\alpha b_i}}{\sigma^2}) m_i}{I_0(\frac{m_i v_0 e^{-\alpha b_i}}{\sigma^2})} e^{-\alpha b_i}, \\[3mm] v_0 \sum_{i=1}^{n} b_i e^{-2\alpha b_i} = \sum_{i=1}^{n} \frac{I_1(\frac{m_i v_0 e^{-\alpha b_i}}{\sigma^2}) m_i}{I_0(\frac{m_i v_0 e^{-\alpha b_i}}{\sigma^2})} b_i e^{-\alpha b_i}. \end{cases}$$

Notice that these score equations differ from (4) only for the Bessel functions ratios $I_1(\frac{m_i v_0 e^{-\alpha b_i}}{\sigma^2}) / I_0(\frac{m_i v_0 e^{-\alpha b_i}}{\sigma^2})$, which multiplies the observations m_i. In particular, this factor decreases the values of observations, since $0 < I_1(x)/I_0(x) < 1$ for $x > 0$, and increases asymptotically to 1 for large SNRs, so that the score equations tend to (4).

As shown in [8], the maximum likelihood estimator for v obtained from an iid sample $M_1, \ldots, M_n \sim \text{Rice}(v, \sigma^2)$ and known σ^2 becomes exactly 0 when the moment estimator for $\mathbb{E}[M^2] = v^2 + 2\sigma^2$ becomes inadmissible, i.e. when $\sum_{i=1}^{n} M_i^2/n - 2\sigma^2 \leq 0$, even if the real value of v is larger than 0. The case of Rice exponential regression suffers from a similar problem in a nontrivial way, and would require σ^2 to be estimated with the other parameters to keep parameter values coherent with the model. Here we will not address this problem, but efforts in this direction are currently in progress.

3.3 Bayesian Approaches

We consider also three different estimators based on a Bayesian posterior distribution: its mean, its median, and its mode. To allow an easy implementation using BUGS code, we introduce a slightly different formulation of the model.

If $M \sim \text{Rice}(v, \sigma^2)$, then $R = M^2/\sigma^2$ has noncentral χ^2 distribution with 2 degrees of freedom and noncentrality parameter $\lambda = v^2/(2\sigma^2)$. Be now $R_1, \ldots, R_n$

the random sample considered, with $R_i = M_i^2/\sigma^2$ and $M_i \sim \text{Rice}(\nu_0\, e^{-\alpha b_i}, \sigma^2)$ for $i = 1, \ldots, n$, and let $\mathbf{r} = (r_1, \ldots, r_n)$ be the observations from this sample. Let $\pi(\nu_0)$ and $\pi(\alpha)$ be the prior distributions of the two unknown parameters, while the density of each R_i will be denoted as $f_{R_i}(r_i)$, with parameter $\lambda_i = \nu_0^2 e^{-2\alpha b_i}/2\sigma^2$. The joint posterior distribution of ν_0 and α is then

$$p(\nu_0, \alpha | \mathbf{r}, \mathbf{b}, \sigma^2) \propto \prod_{i=1}^{n} f_{R_i}(r_i | \lambda_i) \pi(\nu_0) \pi(\alpha).$$

As anticipated in Sect. 2, a noncentral χ^2 distribution of noncentrality λ can be sampled as a mixture of $\chi^2(2P + 2)$ with $P \sim \mathcal{P}(\lambda)$. This allows an easy BUGS implementation of these estimators.

4 Simulation Study

We compared five estimators for α—least squares (LS), maximum likelihood (ML) and posterior mean (PMe), median (PMd) and mode (PMo)—in terms of mean and mean square error. For the two frequentist approaches, ranges for the possible parameter values have been chosen, considering $\nu_0 \in [0.1, 10]$ and $\alpha \in [0.1, 5]$, while the fixed parameter σ^2 has been taken always equal to 1. For the Bayesian point estimators we chose uninformative, uniform priors, with the same support as the ranges chosen for LS and ML. The first two estimators have been computed with R 2.12.2 (see [6]), using built-in optimization functions: optimize for the one-dimensional minimization required in LS and optim, using the L-BFGS-B method, for the likelihood maximization, with startup values $(\nu_{0\text{start}}, \alpha_{\text{start}}) = (1, 1)$. Bayesian posterior distributions have been computed using a Gibbs sampler implemented in JAGS (see [5]). In particular, the following model code (valid for any program supporting BUGS-type language) was used:

```
model{
    for(i in 1:n){
        lambda[i]<-(nu0*nu0)*exp(-2*alpha*b[i])/(2*sigma*sigma)
        p[i]~dpois(lambda[i])
        k[i]<-2*p[i]+2
        M[i]~dchisqr(k[i])
    }
    alpha~dunif(0.1,5)
    nu0~dunif(0.1,10)
}
```

As it can be seen from the model code, uniform prior distributions have been chosen, with supports equal to the search ranges for LS and ML. 10,000 Gibbs sampling iterations have been run for each different sample, with a thinning of 10, and standard diagnostics revealed a good behavior of the generated chains.

We chose b-values in a typical range for diffusion MR machine settings, i.e. from 0 to $1,000 \, \text{s/mm}^2$, on equally spaced grids of $n = 5, 10, 15, 20, 25, 30$ points. Different simulations have been run with parameter values $v_0 = 2, 4, 8$, which represent a low, an intermediate and a high SNR, and $\alpha = 0.7, 1, 3$, typical low, intermediate, and high physiological values of ADC.

It must be reported that the ML estimator, in cases of low SNR, reached the boundaries of the optimization region in various simulations. In the combination $n = 5$, $v_0 = 2$, $\alpha = 3$ only 45% of the simulations gave ML estimates that converged to a value inside the predefined ranges of parameters search, while in the other cases this number oscillated around 70% when $\alpha = 1$ or 100% when $\alpha = 0.7$. These degenerate results have been removed for the computation of bias and variance.

Figure 1 displays the decaying exponential curves we aim to estimate in the nine different combinations of v_0 and α, along with a horizontal line at level σ, to represent the order of magnitude of noise with respect to the signal. The quality of estimates depends both on the SNR at $b = 0$ and on the ADC, as will be clear from simulations.

Figure 2 shows the behavior of bias for the estimators of α with different sample sizes n. For what concerns the frequentist estimators (LS and ML), there is no uniform ordering through the considered values of n when the signal decays slowly ($\alpha = 0.7$), but in the other cases, when noise is stronger along the curve, the maximum likelihood estimate is always less biased than the least squares one; notice also that the least squares estimates do not seem to have a decreasing bias when n increases among the considered values. Concerning the three Bayesian estimators, no striking differences arise among them, while with respect to the frequentist estimators in many cases they have comparable or higher bias, with the exception of the "worst case" $v_0 = 2$, $\alpha = 3$, where they are uniformly more accurate.

From what concerns variance, analyzed in Fig. 3, the LS estimator shows almost always the best performance, excepted for low sample sizes when $\alpha = 3$. The other estimators have similar performances and behaviors at different sample sizes n, with ML and PMe having strikingly higher variance in some noisy cases. As expected, variance notably decreases for all estimators at increasing n in most combinations of parameters, but with very low SNR ($v_0 = 2$) the only one showing empirical convergence of variance to 0 is LS.

An overall index of estimator performance can be evaluated by the mean square error (MSE). Since the MSE is the sum of square bias and variance, the orders of magnitude of these two characteristics assume an important role. As it can be seen from Fig. 4, the LS estimator has the lowest MSE when $\alpha = 0.7, 1$, but exhibits the worst performances in the critical cases of high ADC, where Bayesian estimators seem to work better.

Results for v_0 are not detailed here, but it is worth mentioning that, since it is necessary to estimate the two parameters jointly, the precisions and accuracies of their estimators are mutually influenced. Anyway, estimators for v_0 show a more classical behavior: the LS estimator is in all cases less accurate but more precise (high bias and low variance), and the consistency of all estimators is evident when increasing n. The summary plots for the MSE of the estimators for v_0 can be seen in Fig. 5.

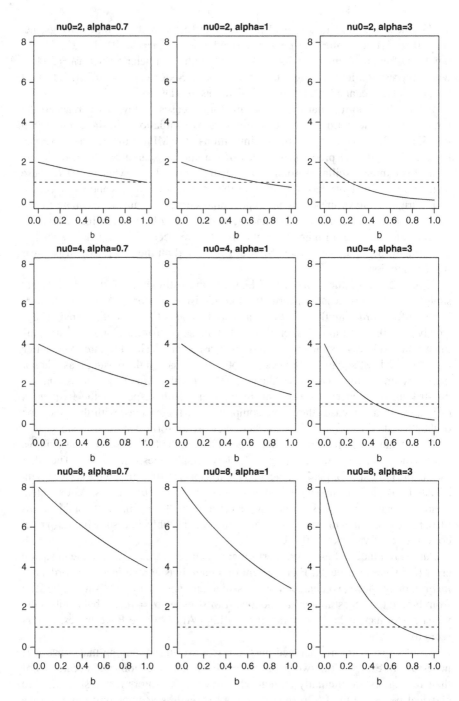

Fig. 1 Stejskal–Tanner model in simulation parameter combinations. b-values are expressed in $1,000\,\text{s/mm}^2$

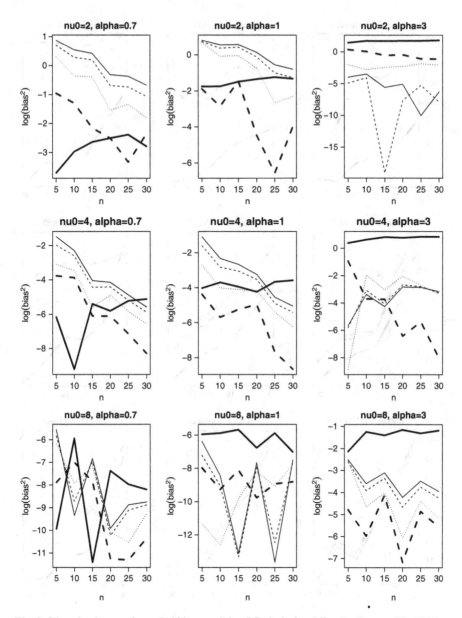

Fig. 2 Bias of estimators for α. *Bold lines*: *solid* = LS, *dashed* = ML; *slim lines*: *solid* = PMe, *dashed* = PMd, *dotted* = PMo

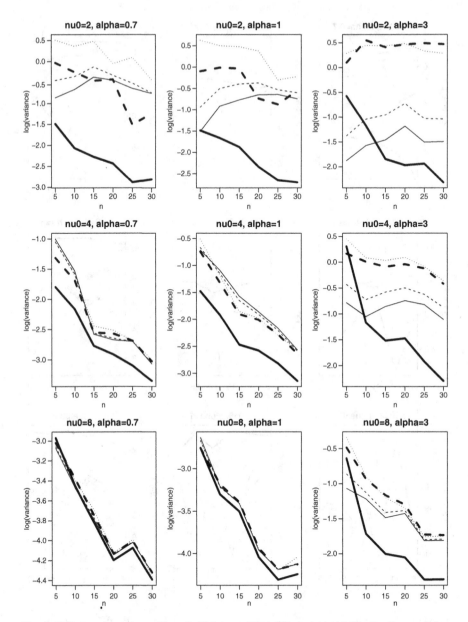

Fig. 3 Variance of estimators for α. *Bold lines*: *solid* = LS, *dashed* = ML; *slim lines*: *solid* = PMe, *dashed* = PMd, *dotted* = PMo

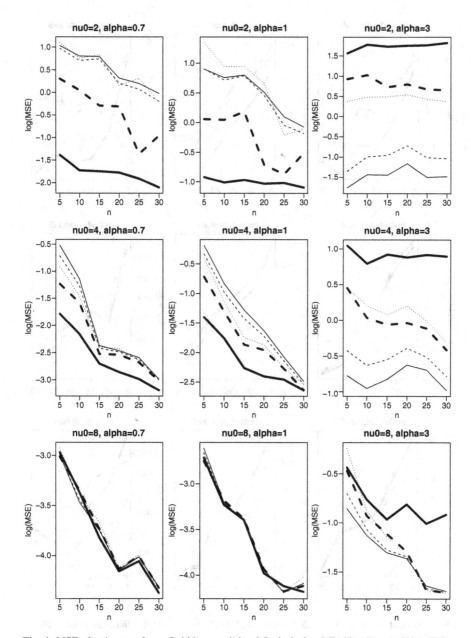

Fig. 4 MSE of estimators for α. *Bold lines*: *solid* = LS, *dashed* = ML; *slim lines*: *solid* = PMe, *dashed* = PMd, *dotted* = PMo

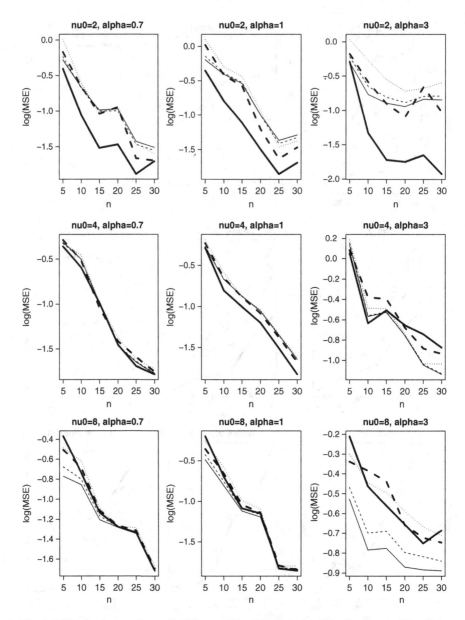

Fig. 5 MSE of estimators for ν_0. *Bold lines*: *solid* = LS, *dashed* = ML; *slim lines*: *solid* = PMe, *dashed* = PMd, *dotted* = PMo

5 Conclusions

In this work, we proposed different methods for estimating pixelwise the ADC from diffusion MR signals, following the Rice noise model and the Stejskal–Tanner equation for magnitude decay. The presented estimators exhibit different features that should be taken into account when approaching real data. The least squares approach is the fastest and has low variance, but becomes less accurate when the conditional signal distribution at different b-values is more distant from normality. The maximum likelihood estimator is slightly slower, requiring a nonlinear maximization on two variables, and has the lowest bias in many cases, but, as pointed out before, it may diverge with samples from noisy signals. Bayesian estimators are the most expensive in terms of computational costs and may require further tuning for improving their performances; they are the best in terms of mean square error at the high ADC here tested and offer the advantage of providing the whole posterior distribution for inferential purposes, while inferential tools regarding LS and ML should rely, at present time, on normal approximations, which may not be reliable with low sample sizes and SNRs. Future studies will focus on inferential aspects, while extending in efficient ways these estimation methods to full MR images. The simultaneous estimation of the dispersion parameter σ^2 will also be developed and tested, requiring some added computational effort to estimation algorithms.

References

1. Abramowitz, M., Stegun, I.A. (eds.): Handbook of Mathematical Functions. Dover, New York (1964)
2. Koay, C.G., Basser, P.J.: Analytically exact correction scheme for signal extraction from noisy magnitude MR signals. J. Magn. Reson. **179**, 317–322 (2006)
3. Landini, L., Positano, V., Santarelli, M.F. (eds.): Advanced Image Processing in Magnetic Resonance Imaging. CRC, West Palm Beach (2005)
4. Lee, M.: BayesSDT: Software for Bayesian inference with signal detection theory. Behav. Res. Methods **40**(3), 450–456 (2008)
5. Plummer, M.: Jags: a program for analysis of Bayesian graphical models using Gibbs sampling. In: Proceedings of the 3rd International Workshop on Distributed Statistical Computing (DSC 2003). Vienna, Austria (2003)
6. R Development Core Team: R: a language and environment for statistical computing. R Foundation for Statistical Computing, Vienna, Austria (2009). URL http://www.r-project.org. Available at: http://cran.r-project.org/, accessed December 2012.
7. Sato, C., Naganawa, S., Nakamura, T., Kumada, H., Miura, S., Takizawa, O., Ishigaki, T.: Differentiation of noncancerous tissue and cancer lesions by apparent diffusion coefficient values in transition and peripheral zones of the prostate. J. Magn. Reson. Imag. **21**(3), 258–262 (2005)
8. Sijbers, J., den Dekker, A.J., Scheunders, P., Van Dyck, D.: Maximum likelihood estimation of Rician distribution parameters. IEEE Trans. Med. Imag. **17**, 357–361 (1998)

9. Walker-Samuel, S., Orton, M., McPhail, L.D., Robinson, S.P.: Robust estimation of the apparent diffusion coefficient (ADC) in heterogeneous solid tumors. Magn. Reson. Med. **62**(2), 420–429 (2009)
10. Woodhams, R., Matsunaga, K., Iwabuchi, K., Kan, S., Hata, H., Kuranami, M., Watanabe, M., Hayakawa, K.: Diffusion-weighted imaging of malignant breast tumors: the usefulness of apparent diffusion coefficient (ADC) value and ADC map for the detection of malignant breast tumors and evaluation of cancer extension. J. Comput. Assist. Tomo. **29**(5), 644–649 (2005)

Longitudinal Patterns of Financial Product Ownership: A Latent Growth Mixture Approach

Francesca Bassi and José G. Dias

Abstract The main goal of this study is to analyze the dynamic process of financial product ownership under the assumption of heterogeneous growth by latent growth mixture models. Using panel data from a survey conducted by the Bank of Italy, we conclude that the trajectory of Italian households in terms of financial product ownership in the period 2000–2006 is homogeneous. Moreover, the process allowed the identification of an outlier trajectory and the obtainment of robust estimates for the population parameters.

1 Introduction

Latent growth models (LGM) consider both intra-individual change and inter-individual differences in such change by estimating the amount of variation across individuals in the latent growth factors (random intercepts and slopes) as well as the average growth [9]. The assumption of homogeneity in the growth parameters—same parameters for all individuals—is not always realistic. If heterogeneity exists and is ignored, statistical results may be seriously biased.

Mixture modeling aims to unmix the population in an unknown number of latent classes or subpopulations [6]. Thus, latent growth mixture modeling (LGMM) allows that the population of interest is not homogeneous, but consisting of subpopulations with varying parameters and within-class variation [11].

F. Bassi (✉)
Department of Statistical Sciences, University of Padua, Padua, Italy
e-mail: francesca.bassi@unipd.it

J.G. Dias
ISCTE–IUL, Instituto Universitário de Lisboa, Lisboa, Portugal
e-mail: jose.dias@iscte.pt

M. Grigoletto et al. (eds.), *Complex Models and Computational Methods in Statistics*,
Contributions to Statistics, DOI 10.1007/978-88-470-2871-5_3,
© Springer-Verlag Italia 2013

This paper illustrates the enormous potential of this type of longitudinal latent variable modeling that combines discrete and continuous latent variables. The application estimates the evolution of financial product ownership at household level in Italy in the period of 2000–2006. We model the binary indicators of ownership (e.g., whether a given household owns a certain type of financial asset) as multiple indicators of a latent process that can differ at segment level. The structure of the paper is as follows: Sect. 2 presents the methodology, Sect. 3 provides the application of the latent growth mixture model to the financial product ownership dataset, and Sect. 4 contains concluding remarks.

2 Latent Growth Mixture Model

Let us define the structure of the data being modeled. Each household (i) at a given time (t) may hold or not the financial product j. Thus, y_{ijt} is a binary variable, where 0 is case a household does not own the financial product at that time, and 1 otherwise.

Let $\tilde{y}_{ijt}$ be a continuous score underlying y_{ijt}, i.e., by defining a threshold v_{jt} it turns out $y_{ijt} = 0$, for $\tilde{y}_{ijt} \leq v_{jt}$, and $y_{ijt} = 1$, otherwise. Let h_{it} be the latent variable or score of household financial ownership at time t. Thus, for each household and time point it results a factor-item response model for each indicator:

$$\tilde{y}_{ijt} = \lambda_{jt} h_{it} + \eta_{ijt}, i = 1, \ldots, n; t = 1, \ldots, T, \tag{1}$$

where the intercept is omitted (given by the threshold), the factor loadings are λ_{jt}, h_{it} is a latent variable, and η_{ijt} is a specific residual. The growth model requires measurement invariance of the factors across time, i.e., the thresholds and factor loadings of the indicators are equal over time—v_j and λ_j—, respectively. Thus, we assume factorial invariance. Moreover, for scaling identification the first loading is fixed at one, $\lambda_1 = 1$, and residual variances are fixed at 1, $\text{Var}(\eta_{ijt}) = 1$, and all covariances are 0. One can set the disturbances associated with the first order factors to be equal over time, imposing homogeneity of error variances. When, as in our case, there is the use of multiple indicators that are abstracted by latent factors, the model is called multivariate LGM or second order latent growth model. This extension results especially useful in marketing research where many relevant phenomena are unobservable and multidimensional (see, as an example, [16]).

The latent process h_{it} defines the trajectory of the household financial ownership at time t and is given by

$$h_{it} = h_i^{\text{I}} + (t - 1)h_i^{\text{S}} + \varepsilon_{it}, \tag{2}$$

where h_i^{I} and h_i^{S} are the intercept and slope of the process, respectively, and ε_{it} is the error term of the process. LGMs can be described also in the context of Structural Equation Modeling, in this case, (1) and (2) represent the measurement and the structural component, respectively [8].

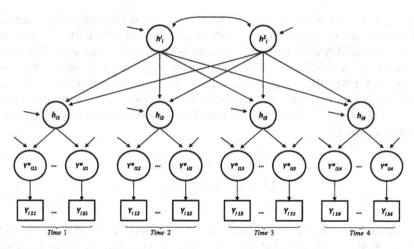

Fig. 1 LGM path diagram

For the conditional model, the intercept and the slope are function of the observed variables contained in the vector $\mathbf{x}_i$ [13]:

$$h_i^{\mathrm{I}} = \alpha_0^{\mathrm{I}} + \boldsymbol{\alpha}_1^{\mathrm{I}} \mathbf{x}_i + \zeta_i^{\mathrm{I}}$$
$$h_i^{\mathrm{S}} = \alpha_0^{\mathrm{S}} + \boldsymbol{\alpha}_1^{\mathrm{S}} \mathbf{x}_i + \zeta_i^{\mathrm{S}},$$

$$(3)$$

where h_i^{I} and h_i^{S} are the continuous latent variables (intercept and slope), α_0^{I} and α_0^{S} are the constants (or means in the case of the unconditional model), and $\boldsymbol{\alpha}_1^{\mathrm{I}}$ and $\boldsymbol{\alpha}_1^{\mathrm{S}}$ contain the coefficients of the covariates in the conditional model.

The disturbances are given by:

$$\zeta = \begin{bmatrix} \zeta_i^{\mathrm{I}} \\ \zeta_i^{\mathrm{S}} \end{bmatrix} \sim N\left(\begin{bmatrix} 0 \\ 0 \end{bmatrix}, \begin{bmatrix} \psi_{\mathrm{II}} & \psi_{\mathrm{IS}} \\ \psi_{\mathrm{IS}} & \psi_{\mathrm{SS}} \end{bmatrix} \right).$$

$$(4)$$

Figure 1 contains the path diagram of an unconditional second order LGM for four time points, in each occasion J indicators (Y_{ijt}) of a latent factor are observed on each unit. The trajectory across latent factors (h_{it}) is described by an intercept (h_i^{I}) and a slope (h_i^{S}) [4].

The latent growth mixture model (LGMM) assumes that the population is heterogeneous, and different subpopulations are characterized by different trajectories [5]. In this case, the model can be defined as a finite mixture of K functions with density

$$f(\mathbf{y}_i; \boldsymbol{\varphi}) = \sum_{k=1}^{K} \pi_k f^{(k)}(\mathbf{y}_i; \boldsymbol{\varphi}_k),$$

$$(5)$$

where φ is the set of all parameters in the model. This mixture model assumes a discrete latent variable with K classes (multinomial distribution), where π_k is the prior probability that individual i belongs to latent class k and φ_k contains the parameters of the conditional distribution. The density function in latent class k is $f^{(k)}(\mathbf{y}_i; \varphi_k)$, where the observed data vector $\mathbf{y}_i$ conditional on the latent class k is connected to $\tilde{\mathbf{y}}_i$ through the threshold parameters. We assume that the measurement model is class-invariant. However, different classes can be characterized by different latent processes, i.e.,

$$h_i^{\mathrm{I}} = \alpha_{0k}^{\mathrm{I}} + \boldsymbol{\alpha}_{1k}^{\mathrm{I}} \mathbf{x}_i + \zeta_{ik}^{\mathrm{I}}$$
$$h_i^{\mathrm{S}} = \alpha_{0k}^{\mathrm{S}} + \boldsymbol{\alpha}_{1k}^{\mathrm{S}} \mathbf{x}_i + \zeta_{ik}^{\mathrm{S}} \tag{6}$$

as described before. Moreover, we have the residuals $\boldsymbol{\varepsilon}^{(k)} \sim N(\mathbf{0}, \boldsymbol{\Theta}^{(k)})$ and $\boldsymbol{\zeta}^{(k)} \sim N(\mathbf{0}, \boldsymbol{\Psi}^{(k)})$ both assumed to be uncorrelated with other variables and with each other. The covariance structures are:

$$\boldsymbol{\Theta}^{(k)} = \mathrm{COV}(\boldsymbol{\varepsilon}^{(k)}) = \mathrm{diag}(\theta_1^{(k)}, \theta_2^{(k)}, \ldots, \theta_T^{(k)})$$

$$\boldsymbol{\Psi}^{(k)} = \mathrm{COV}(\boldsymbol{\zeta}^{(k)}) = \left[\begin{array}{cc} \psi_{\mathrm{II}}^{(k)} & \psi_{\mathrm{IS}}^{(k)} \\ \psi_{\mathrm{SI}}^{(k)} & \psi_{\mathrm{SS}}^{(k)} \end{array} \right].$$

The LGMM approach allows to capture different developmental processes of group membership that cannot be determined a priori [10] and can be understood within the context of latent class models [3] in which the total population can be partitioned into distinct but unobserved sub-populations. Again this extension results appropriate to marketing research to analyze segmented markets [15].

Parameters are estimated by the maximum likelihood (ML) method using the EM algorithm. Estimation consists of two parts: estimation of the latent class sub-model and the estimation of parameters related to the LGM conditional on the latent class sub-model [12, 14]. The log-likelihood function of observed data for the LGMM is

$$\log L(\boldsymbol{\varphi}; \mathbf{y}_i) = \sum_{i=1}^{n} \log L_i(\boldsymbol{\varphi}; \mathbf{y}_i) = \sum_{i=1}^{n} \log f(\mathbf{y}_i; \boldsymbol{\varphi}), \tag{7}$$

where φ is the set of all parameters in the model.

Given that the latent variables are continuous and the manifest variables are binary, we have to integrate out the latent continuous variables in the estimation process using adaptive quadrature. Model selection is based on information criteria (Akaike's information criterion and Bayesian information criterion). In model estimation, we allow a maximum of 1,500 iterations and relative log-likelihood change with a convergence tolerance of 10^{-6}. Given the multimodal log-likelihood function of the mixture model, 20 different random starting values were run in order to avoid local maxima for the two-latent class model. The syntax of the model was written in MPlus 5.

3 The Empirical Study

The Bank of Italy (BI) has been running the Survey of Household Income and Wealth (SHIW) since 1965. With a few exceptions, the survey was conducted on a 2-year basis since then. The reference population is the resident non-institutional population of the country. SHIW provides information on income, savings, consumption expenditure, and the real wealth of Italian households, as well as on household composition and on labor force participation. The questionnaire is divided into several sections. Questions concerning the whole household are answered by the head of household, questions on individual incomes are answered by each income recipient. In 1989, BI introduced a longitudinal component in the survey, adopting a peculiar split panel design: at each survey round (or simply wave, as it is often called), the sample consists of two sections: a panel sub-sample, made up of households who participated in the previous waves; and a fresh cross-sectional sub-sample. Once a household has entered the panel, it stays in the sub-sample till it leaves it because of attrition [7]. Table 1 contains the size of the various samples and sub-samples, fresh and panel from 2000 to 2006.

In this paper we consider the sample of the 1951 panel households interviewed in all the waves from 2000 to 2006. We dispose of information on the ownership of a series of financial products and on family characteristics. The SHIW contains a detailed information on the distribution of Italian households financial assets in 18 categories, such as checking accounts, savings accounts, certificates of deposits, postal deposits, postal bonds, Treasury bills up to 1-year maturity (BOT), floating rate Treasury certificates indexed to BOTs (CCT), fixed-rate long-term Treasury bonds (BTP), other government papers, corporate bonds, investment funds, and equities. Many of these products are hold by a very small percentage of families, so we aggregate this information considering ownership by the household of at least one bank or postal deposit account (PDEP), at least one bank or postal savings account (PLIB), of shares (SHARES), state bonds (SBONDS), and other types of bonds (BONDS).

Table 2 lists the percentage of households owning the considered financial products in the 4 years of analysis. Region of the country where living (REGION) and gender of the head of household (HEAD) are added as time constant covariates (observed value in 2000), making the modeling of the intercept and slope of the latent process conditional. In 2000, 46.0% of the sample households were from the North of Italy, and 21.3% and 32.7% were from the Centre and the South, respectively. Moreover, in 2000, 29.8% of the panel households were headed by a woman.

Preliminary analysis by the Bank of Italy [2] showed that financial product ownership by Italian households varies with income, education, and professional condition of the head of household, and area of the country where the family lives. Albareto et al. [1], using data from the SHIW, show that the composition of financial wealth of Italian households underwent a remarkable change during the period 1998–2008, as it can be inferred also from the data in Table 2. For example, the

Table 1 Italian survey on household income and wealth survey plan, 2000–2006 and number of interviewed families per year and wave

Year of first	Waves			
interview	2000	2002	2004	2006
1987	61	44	33	30
1989	343	263	197	159
1991	832	613	464	393
1993	399	270	199	157
1995	245	177	117	101
1998	1,993	1,224	845	636
2000	4,128	1,014	667	475
2002		4,406	1,082	672
2004			4,408	1,334
2006				3,811
Total	8,001	8,011	8,012	7,768
Panel proportion %	48.4	45.0	45.0	50.9

Table 2 Dependent variables—Percentages

	2000	2002	2004	2006
Bank and postal deposit accounts (*PDEP*)	83.6	84.6	84.7	88.6
Bank and postal saving accounts (*PLIB*)	22.9	22.8	21.8	20.5
Shares (*SHARES*)	14.9	13.6	12.0	11.1
State Bonds (*SBONDS*)	15.8	12.7	10.8	11.4
Bonds (*BONDS*)	10.6	11.1	11.6	11.3

percentage of families owning deposits accounts increased over the period, while that of those owning saving accounts diminished and the diffusion of state bonds and shares declined. Our analysis wants to understand more in-depth these dynamics.

To assess heterogeneity in the household panel sample, the first step is to select the number of latent classes needed to describe the dynamic process, i.e., to check whether a model with two classes (heterogeneity) performs better than a model with only one class (homogeneity). Both the Bayesian Information Criterion (BIC) and the Akaike Information Criterion (AIC) indexes suggest that a model with one latent class fits better the data than the model with two latent classes. As BIC and AIC are monotonic functions of the number of latent classes, there is no need to estimate models with more than two classes as the BIC and AIC will be always worse than the two-latent class model. Tables 3 and 4 list structural and measurement estimated model parameters, respectively.

We assume that measurement model parameters—loadings and threshold—are class and time invariant. We observe that the two-latent class model tends to increase loadings estimates. This is particularly true for the residual variances estimates that partially explains the lack of fit of the second model.

The main focus of the analysis is on the structural part of the model, i.e., the latent trajectories measured by the indicators and explained by the covariates

Table 3 Estimated parameters—measurement component—one and two latent classes

	One latent class		Two latent classes	
	Estimate	S.E.	Estimate	S.E.
Loadings (λ_j)				
PDEP	1	–	1	–
PLIB	−0.07	0.03	−0.11	0.06
SHARES	1.30	0.30	3.98	1.39
BONDS	0.76	0.11	1.47	0.13
SBONDS	0.64	0.09	0.93	<0.01
Thresholds (ν_j)				
PDEP	−3.17	0.20	−1.55	0.11
PLIB	1.28	0.04	1.21	0.05
SHARES	3.61	0.28	7.32	1.57
BONDS	2.82	0.23	3.67	0.23
SBONDS	2.45	0.21	2.76	<0.01
Residual variances (Θ)				
θ_1	0.08	0.01	0.24	0.03
θ_2	<0.01	<0.001	0.18	0.04
θ_3	<0.01	<0.01	0.19	0.04
θ_4	0.08	<0.01	0.24	0.04

Table 4 Estimated parameters—structural component—one- and two-latent classes

	One latent class		Two latent classes			
			Class 1		Class 2	
	Estimate	S.E.	Estimate	S.E.	Estimate	S.E.
Class proportions	1	–	0.95	–	0.05	–
Intercept (α^I)						
Constant	0	–	0.64	0.15	0	–
REG_N	0.46	0.26	0.22	0.12	2.04	0.02
REG_S	−1.84	0.20	−0.76	0.13	−25.18	0.06
SEX	0.46	0.20	0.25	0.11	25.30	0.06
Slope (α^S)						
Constant	−0.15	0.07	−0.14	0.08	301.62	<0.01
REG_N	0.09	0.05	0.03	0.04	−301.59	0.01
REG_S	−0.13	0.08	−0.10	0.08	−292.61	0.02
SEX	0.16	0.05	0.13	0.04	−8.82	0.01
Covariance (Ψ)						
ψ_{II}	3.35	0.63	0.66	0.094	0.66	0.09
ψ_{SS}	0.11	0.03	0.01	0.008	0.01	<0.01
ψ_{IS}	−0.09	0.04	−0.05	0.04	−0.13	0.03

within each latent class (Table 4). Looking at estimated class proportions, we may conclude that the second latent class is a residual one, containing only a 5% of households and describing an outlier trajectory in financial products ownership. Therefore, estimated parameters for latent class one, in the two-latent class model, can be interpreted as a robust estimation of the single trajectory, characterizing the entire sample of households, obtained removing the impact of outlier observations, identified by latent class two.

The structural component of the model postulates that covariates REGION (area of the country where the household lives) and HEAD (gender of head of household) have an influence on the latent factors—intercept and slope. Looking at estimated parameters, we conclude that:

- The constant parameter in the trajectory intercept is statistically significant, indicating that there is a significant level of financial product ownership in year 2000.
- The constant parameter in the trajectory slope is negative but not statistically significant, showing a constant trend in the ownership of financial products by Italian households.
- Estimated regression coefficients for the covariate REGION show a significant effect of the fact that the household lives in the South of Italy in year 2000 (*REG_S*), being "living in the Centre" the reference category; this effect, however, is not significantly different from that of living in the North of the country. Living in the South reduces the financial score of 0.76 at the beginning of the observational period (year 2000).
- The fact that the head of household is a man has a significant and positive impact on the intercept of the trajectory.
- For what concerns, the trajectory slope, the impact of living in the South of the country is no longer significant, while the fact that a man is the head of household has a positive and significant impact.

We have restricted the covariance matrices $\boldsymbol{\Theta}^{(k)}$ and $\boldsymbol{\Psi}^{(k)}$ to be class invariant. Analyzing the estimates of the elements of matrix $\boldsymbol{\Psi}$, we may conclude that the variance of the intercept is significantly different from 0 whilst this is not the case for the slope. This result indicates the existence of substantial variation among households in the initial condition in year 2000 that is not taken into account by the covariates included in the conditional model. The estimated covariance between the two factors, intercept and slope, is not statistically significant, indicating that there is no relation between the two, i.e., the level of financial product ownership in year 2000 is not related to the growth in the following period. This conclusion is true for the single-latent class model and for latent class one in the two-latent class model, which contains the robust estimates.

Using posterior estimated probabilities, it is possible to identify the outlier households. Table 5 contains an example on outlier patterns: with a probability of 0.555, a household with 1 female component living in the South of the country shows such a pattern of financial product ownership.

Table 5 Observed financial product ownership with probability 0.555 for a household with 1 female component living in the South

	PDEP	PLIB	SHARES	SBONDS	BONDS
2000	0	1	0	0	0
2002	0	1	0	0	0
2004	0	1	0	0	0
2006	0	1	0	0	0

Table 6 Estimated parameters of one latent class model (after removing outliers)

	Loadings (λ_j)			Thresholds (τ_j)	
	Estimate	S.E.		Estimate	S.E.
PDEP	1	–		−3.16	0.20
PLIB	−0.07	0.03		1.29	0.04
SHARES	1.40	0.34		3.55	0.28
SBONDS	0.80	0.12		2.77	0.25
BONDS	0.67	0.09		2.40	0.22
	Intercept (α^I)			Slope (α^S)	
	Estimate	S.E.		Estimate	S.E.
constant	0	–		−0.18	0.07
REG_N	0.39	0.26		0.08	0.05
REG_S	−1.73	0.19		−0.15	0.08
HEAD	0.40	0.19		0.17	0.04
	Θ			Ψ	
	Estimate	S.E.		Estimate	S.E.
θ_1	0.09	0.01	ψ_{II}	2.86	0.56
θ_2	<0.01	<0.01	ψ_{SS}	0.10	0.03
θ_3	<0.01	<0.01	ψ_{IS}	−0.04	0.04
θ_4	0.17	0.02			

After removing the outlier households, we reestimated the one-class model obtaining the results listed in Table 6.

4 Conclusions

This paper estimates a second order latent growth mixture model to answer two important research questions in longitudinal data analysis that cannot be addressed by the conventional latent growth model. Combining continuous and discrete latent variables, it identifies unobserved sub-populations with different trajectories or dynamics. The findings of this study demonstrate the existence of a single longitudinal change pattern for the Italian households in this period. The model is able to identify specific household patterns that are classified as outliers. From the robust results (latent class 1 in the two-latent class model), we identify that gender of head of household explains both the initial level and the trajectory of the financial score obtained from the observed ownership patterns of five financial products available to households whereas living in a specific area of the country (North,

Centre, South) is important to measure the initial level of the dynamic process. The results are confirmed after excluding the outlier households and reestimating a single latent growth model.

Acknowledgments Research for this paper was supported by the project financed by the Italian Ministry of University and Education PRIN 2009 with title "The influence of social contexts and consumption circumstances in the perception of cognitive age in older consumers: assessment and new statistical methods of measurement" and Fundação para a Ciência e a Tecnologia (Portugal) Grant PTDC/CS-DEM/108033/2008.

References

1. Albareto, G., Bronzini, R., Caparra, D., Carmignani, A., Venturini, A.: The real and financial wealth of Italian household by region. In: Household Wealth in Italy, pp. 57–77. Banca d'Italia, Roma (2008)
2. Banca, d'Italia: Supplementi al Bollettino Statistico. I bilanci delle famiglie italiane 2006, Roma (2008)
3. Clogg, C.C.: In: Arminger, A., Clogg, C.C., Sobel, M.E. (eds.) Latent Class Models. Handbook of Statistical Modeling for the Social and Behavioural Sciences, Chap. 6, pp. 311–359. Plenum, New York (1995)
4. Collins, L.M., Sayer, A.G. (eds.): New Methods for the Analysis of Change. American Psychological Association, Washington (2001)
5. Connell, A.M., Frye, A.A.: Growth mixture modelling in developmental psychology: overview and demonstration of heterogeneity in developmental trajectories of adolescent antisocial behaviour. Infant Child Develop. **15**, 609–621 (2006)
6. Duncan, T.E., Susan, S.C., Strycker, L.A., Okut, H.: Growth mixture modelling of adolescent alcohol use data. Chapter Addendum to an Introduction to Latent Variable Growth Curve Modeling: Concepts Issues, and Applications. Oregon Research Institute, Eugene (2002)
7. Giraldo, A., Rettore, E., Trivellato, A.: Attrition bias in the Bank of Italy's survey of household income and wealth. Working Paper 41, Prin "Occupazione e disoccupazione in Italia: misura e analisi dei comportamenti", Department of Statistics, University of Padova, Italy (2001)
8. Ghisletta, P., McArdle, J.J.: Latent growth curve analyses of the development of height. Struct. Equ. Model. **8**, 531–555 (2001)
9. Kreuter, F., Muthén, B.: Analyzing criminal trajectory profiles: bridging multilevel and group-based approaches using growth mixture modelling. J. Quantit. Criminol. **24**, 1–31 (2008)
10. Li, F., Duncan, T.E., Duncan, S.C.: Latent growth modelling of longitudinal data: a finite growth mixture modeling approach. Struct. Equ. Model. **8**, 493–530 (2001)
11. Muthén, B.: The potential of growth mixture modeling. Commentary. Infant Child Develop. **15**, 623–625 (2006)
12. Muthén, B., Shedden, K.: Finite mixture modelling with mixture outcomes using the EM algorithm. Biometrics **55**, 463–469 (1999)
13. Salgueiro, M.F., Smith, P.W.F., Vieira, M.T.D.: A multi-process second-order growth curve model for subjective well-being. Quality Quantity 1–18 (2011) Doi: 10.1007/s11135-011-9541-y
14. Yung, Y.F.: Finite mixtures in confirmatory factor-analysis models. Psychometrika **62**, 297–330 (1997)
15. Wedel, M., Kamakura, W.A.: Market Segmentation: Concepts and Methodological Foundations. Kluwer, Boston (2000)
16. Wiesmayer, C.: Longitudinal satisfaction measurement using latent growth curve models and extensions. J. Retailing Consum. Serv. **17**, 321–331 (2010)

Computationally Efficient Inference Procedures for Vast Dimensional Realized Covariance Models

Luc Bauwens and Giuseppe Storti

Abstract This paper illustrates some computationally efficient estimation procedures for the estimation of vast dimensional realized covariance models. In particular, we derive a Composite Maximum Likelihood (CML) estimator for the parameters of a Conditionally Autoregressive Wishart (CAW) model incorporating scalar system matrices and covariance targeting. The finite sample statistical properties of this estimator are investigated by means of a Monte Carlo simulation study in which the data generating process is assumed to be given by a scalar CAW model. The performance of the CML estimator is satisfactory in all the settings considered although a relevant finding of our study is that the efficiency of the CML estimator is critically dependent on the implementation settings chosen by modeller and, more specifically, on the dimension of the marginal log-likelihoods used to build the composite likelihood functions.

1 Introduction

Many financial applications, such as portfolio optimization or risk management, require to work with large dimensional portfolios involving a number of assets in the order of 100 or even more. In order to obtain parsimonious multivariate volatility models, whose estimation is feasible in large dimensions, it is necessary to impose drastic homogeneity restrictions on the dynamic laws determining the evolution of conditional variances and covariances. However, even for parsimonious models, for very large dimensions the computation of (quasi) maximum likelihood estimates can

L. Bauwens
CORE, Université Catholique de Louvain, Louvain-la-Neuve, Belgium
e-mail: luc.bauwens@uclouvain.be

G. Storti (✉)
DiSES, University of Salerno, Fisciano (SA), Italy
e-mail: storti@unisa.it

M. Grigoletto et al. (eds.), *Complex Models and Computational Methods in Statistics*, Contributions to Statistics, DOI 10.1007/978-88-470-2871-5_4,
© Springer-Verlag Italia 2013

be computationally challenging and troublesome. Also it is important to note that for moderately large values of the model's dimension n (as a rule of the thumb, say for $50 \leq n \leq 100$), even if direct maximization of the likelihood or quasi likelihood function is feasible, the computational time required can be so high to prevent the use of resampling and simulation techniques and, in general, any application that implies the need to iteratively estimate the model for a large number of times. The application of bootstrap to realized covariance models has been recently investigated by [5] while the application of simulation-based inference procedures is usually required for long-term prediction of risk measures such as Value at Risk and Expected Shortfall.

This has stimulated the research on the development of alternative algorithms for the generation of computationally efficient consistent estimators of the parameters of vast dimensional multivariate volatility models. These algorithms have been first developed for the estimation of Multivariate GARCH (MGARCH) models, see [2] for a review, but they can be modified or adapted, as shown in [3], for their application to models for realized covariance matrices.

An obvious approach to deal with inference in vast dimensional systems is to split the multivariate estimation problem into a set of simpler low-dimensional problems. This is the spirit of the McGyver method proposed by [7]. The algorithm is illustrated for the case of Dynamic Conditional Correlation (DCC) models with correlation targeting [6] although it can be readily applied to any *scalar* MGARCH model such as a scalar BEKK model [8][1]. The basic idea is that, if the process dynamics are characterized by scalar parameters, these can be consistently estimated by fitting the model even to an arbitrarily chosen bivariate subsystem. The estimation can then be repeated over all the possible $n(n-1)/2$ different bivariate subsystems or a subset of them. The final estimate of model parameters is obtained by calculating the mean or median of the empirical distribution of the estimated coefficients. This estimate is expected to be more efficient than an estimate obtained by fitting the model to a single bivariate subsystem since it is implicitly pooling information from all the assets in the dataset. Evidence in this direction is provided by [7]. The procedure is in its nature heuristic but it is straightforward to show that it automatically returns a consistent estimator. However, [7] does not provide any analytical results on the asymptotic properties of the estimator, including its asymptotic distribution and efficiency. On the other hand, the finite sample statistical properties of the McGyver estimator are investigated by Monte Carlo simulation. One point which is left unexplored by [7] is related to the sensitivity of the properties of the estimation procedure to the size of the subsystems involved in its implementation.

An alternative approach to the estimation of vast dimensional conditional heteroskedastic models is based on Composite Likelihood theory (see [14] for a recent review). This approach replaces the full log-likelihood with an approximation

---

[1]We indicate as *scalar* any model in which all the conditional covariances or correlations share the same parameters.

based on the sum of low-dimensional log-likelihoods of a given dimension $m << n$. As for the McGyver method, the researcher can consider the full set of m-dimensional log-likelihoods or a subset of these. In [9] the CML approach is applied to the estimation of scalar DCC models with correlation targeting. From a computational point of view, the calculation of CML estimators is much faster than that of standard ML estimators. Reference [9] also analyzes different variants of the Composite Maximum Likelihood (CML) estimator and assesses their finite sample properties by means of an extensive Monte Carlo study. Their findings show that, in large systems, the CML estimator can be much more efficient than the Maximum Likelihood (ML) estimator with correlation targeting. In particular, the simulation results reveal that this estimator tends to be affected by a systematic bias component whose size is increasing with n. In the paper this difference is ascribed to the high number of nuisance parameters involved in the optimization. Also, they show that the CML estimator favorably compares with the McGyver method being, by far, more efficient than the latter. In the light of these results, in this paper our attention will be focused on CML estimation.

Both the CML and the McGyver method reformulate the estimation problem in terms of simpler lower dimensional problems but while, in the McGyver method, the final estimate is obtained as a function of the results of several low-dimensional optimizations, in CML estimation the optimization is performed just once leading to a substantial reduction of the computing time. Except for the recent contribution by [3], to our knowledge, so far this inference procedure has not been applied to the estimation of models for realized covariances. In this paper we will illustrate an approach to the estimation of large dimensional realized covariance models based on the maximization of a CL function derived under the assumption of Wishart marginal log-likelihoods. By means of a Monte Carlo simulation study we will (1) evaluate the efficiency of the estimator in finite samples (2) investigate the sensitivity of the estimator's performance, bias and efficiency, to the size of the marginal log-likelihoods used to build the CL function and to the number of low-dimensional subsystems used in the computation.

Among the several different models for realized covariance matrices that have been proposed in the past years, we will focus on the class of Conditional Autoregressive Wishart (CAW) models recently introduced by [10]. These models are based on the assumption that the conditional distribution of the realized covariance matrix is a Wishart distribution with time-varying conditional expectation proportional to the scale matrix of the Wishart. In the basic version of the model the time-varying conditional expectation of the realized covariance matrix is assumed to follow a BEKK [8] type specification. Unless further restrictions on the parameter space are imposed, this assumption still leaves the number of parameters linear in n^2 where n is the number of assets considered. In [10] the authors present an application to a dataset including 5 stocks where the estimated model includes 116 parameters. It is so easy to understand how fitting an unconstrained CAW model to a dataset whose dimension is even moderately large is not feasible. Hence, in this paper, we will consider a restricted version of the CAW model in which the parameter matrices of the dynamic equation for the conditional expectation of the realized covariance matrix are assumed to be scalar. A covariance targeting approach will be also used in order to avoid direct estimation of the matrix of intercepts of the BEKK recursion.

The structure of the paper is as follows. In Sect. 2 we will illustrate the CAW model and discuss its statistical properties. Maximum likelihood inference for the CAW model will be discussed in Sect. 3 while, in Sect. 4, we will present the CML estimator for the parameters of a restricted CAW model. Section 5 will report the results of a Monte Carlo simulation study and conclude.

2 The Conditional Autoregressive Wishart Model

The CAW model, recently proposed by [10], is based on the assumption that the conditional distribution of realized covariance matrices, given past information, is Wishart. In the literature on realized covariance models the Wishart assumption is not new and has already been used in a number of papers, starting from [11] who were probably the first to use the idea of a Wishart process for realized covariance matrices with the WAR(p) (Wishart autoregressive) process, where p is a lag order parameter. They assume that the conditional distribution of realized covariance matrices follows a non-central Wishart distribution where the matrix of non-centrality parameters is modeled as a function of past lagged realized covariance matrices. Practical application of this model has been limited by the fact that it is too heavily parameterized since it uses a number of parameters equal to $3n^2/2 + n/2 + 1$ (for the one lag case). In order to obtain a more parsimonious model structure, [4] proposes a block structured version of the WAR model which significantly reduces the number of parameters but still keeps it on the order of n^2. A different approach is taken by [12, 13] who propose to jointly model the returns vector and its realized covariance matrix. In both papers, just as in the CAW model, the realized covariance part of the model specifies the conditional distribution of the realized covariance as a Wishart, whose expected value is proportional to the scale matrix of the Wishart. In [12], that scale matrix is modelled as a function of a few lags of itself while, in [13], it is assumed to follow a BEKK process. The CAW model shares with [13] the assumption of conditional Wishart distribution and BEKK-type dynamics for the realized covariance models.

More specifically, let C_t, for $t = 1, \ldots, T$, be a time series of positive definite symmetric (PDS) realized covariance matrices of dimension n. It is assumed that the conditional distribution of C_t, given past information on the history of the process I_{t-1}, consisting of C_τ for $\tau \leq t - 1$, and $\forall t$, is given by a n-dimensional central Wishart distribution

$$C_t | I_{t-1} \sim W_n(\nu, S_t/\nu), \tag{1}$$

where ν ($> n - 1$) is the degrees of freedom parameter and S_t/ν is a PDS scale matrix of order n. From the properties of the Wishart distribution (see [1], among the others), it follows that

$$E(C_t | I_{t-1}) = S_t, \tag{2}$$

so that the i, j-th element of S_t is defined as the conditional covariance between returns on assets i and j, $\text{cov}(r_{i,t}, r_{j,t} | I_{t-1})$, for $i, j = 1, \ldots, n$, $r_{i,t}$ denoting the logarithmic return on asset i between the ends of periods $t - 1$ and t. Equation (1) defines a generic CAW model as proposed by [10]. The specification of the dynamic updating equation for S_t can be chosen within a wide range of options. Namely [10] uses a BEKK-type formulation mutuated from the MGARCH literature. For a model of order (p, q), this corresponds to the following dynamic equation

$$S_t = GG' + \sum_{i=1}^{p} A_i C_{t-i} A'_i + \sum_{j=1}^{q} B_j S_{t-j} B'_j \qquad (3)$$

where A_i and B_j are square matrices of order n and G is a lower triangular matrix such that GG' is PDS. This choice also ensures that S_t is PDS for all t if S_0 is itself PDS. In order to guarantee model identifiability, it is sufficient to assume that the main diagonal elements of G and the first diagonal element for each of the matrices A_i, B_j are positive. There are two main differences between the WAR model of [11] and the CAW model: (1) it is assumed that the conditional distribution of C_t is a central Wishart distribution rather than a non-central one (2) the CAW model analyzes the dynamic evolution of the scale matrix S_t while the WAR model is focusing on the matrix of non-centrality parameters.

In order to investigate the statistical properties of the CAW(p, q) model, it is useful to consider two alternative observationally equivalent representations of the model.

First, the CAW(p, q) model in (1–3) can be represented as a state-space model with observation equation given by

$$C_t = \frac{1}{\nu} S_t^{1/2} U_t (S_t^{1/2})' \qquad U_t \sim W_n(\nu, I_n); \qquad (4)$$

where $S_t^{1/2}$ denotes the lower triangular matrix obtained from the Cholesky factorization of S_t such that $S_t^{1/2}(S_t^{1/2})' = S_t$ and U_t is a measurement error distributed as a standardized Wishart distribution with identity scale matrix and degrees of freedom equal to ν. S_t acts as a matrix-variate state variable with dynamic transition equation given by (3). This representation allows to interpret S_t as the "true" latent integrated covariance for a wide class of multivariate continuous time stochastic volatility models, C_t as a consistent estimator of S_t and U_t as the associated matrix of estimation errors. Second, a CAW(p, q) process admits the following VARMA representation

$$c_t = g + \sum_{i=1}^{\max(p,q)} (\mathcal{A}_i + \mathcal{B}_i) c_{t-i} + \sum_{j=1}^{q} \mathcal{B}_j v_{t-j} + v_t \qquad (5)$$

where $g = \text{vech}(GG')$ and $c_t = \text{vech}(C_t)$; v_t is a martingale difference error term such that $E(v_t) = 0$ and $E(v_t v'_s) = 0$, $\forall t \neq s$; $\mathcal{A}_i = L_{n*}(A_i \otimes A_i) D_{n*}$ and

$\mathcal{B}_j = L_{n*}(B_j \otimes B_j)D_{n*}$, with $n* = n(n+1)/2$ ($\mathcal{A}_i = 0$ for $i > p$, $\mathcal{B}_j = 0$ for $j > q$). L_{n*} and D_{n*} are duplication and elimination matrices such that $\text{vec}(X) = D_{n*}\text{vech}(X)$ and $\text{vech}(X) = L_{n*}\text{vec}(X)$. This representation allows to derive conditions for the existence of the unconditional mean of the CAW(p,q) process. In particular, it can be shown that $\bar{c} = E(c_t)$ will be finite if and only if all the eigenvalues of the matrix $\Psi = \mathcal{A}_i + \mathcal{B}_j$ are less than 1 in modulus. In this case we will have

$$\bar{c} = E(c_t) = \left(I_{n*} - \sum_{i=1}^{\max(p,q)} (\mathcal{A}_i + \mathcal{B}_i) \right)^{-1} g. \tag{6}$$

For large n, the model formulation in (3) renders the estimation unfeasible due to the high number of parameters. As we are interested in applying the model to situations in which n is large (say $n \approx 50$), we consider a modified version of (3) in which the parameter matrices A_i and B_j are replaced by scalars and covariance targeting is used to avoid simultaneous estimation of the matrix G

$$S_t = \left(1 - \sum_{i=1}^{p} \alpha_i - \sum_{j=1}^{q} \beta_j \right) \bar{C} + \sum_{i=1}^{p} \alpha_i C_{t-i} + \sum_{j=1}^{q} \beta_j S_{t-j} \tag{7}$$

with $\bar{C} = E(C_t)$, $\alpha_i = a_i^2$ and $\beta_j = b_j^2$. The particular implementation of covariance targeting used in this formulation is justified by the fact that applying equation (6) to model (7) returns

$$\bar{C} = \frac{GG'}{(1 - \sum_{i=1}^{p}\alpha_i - \sum_{j=1}^{q}\beta_j)}.$$

In practice, covariance targeting is implemented by consistently pre-estimating the unconditional expectation of C_t by the sample average $\hat{\bar{C}} = T^{-1}\sum_{t=1}^{T} C_t$ and substituting this estimator for the corresponding population moment in (7). The scalar model implies that the conditional variances and covariances all follow the same dynamic pattern which is indeed a restrictive assumption. However, this compromise is necessary in order to obtain a parsimonious model whose estimation is tractable in high dimensions. A generalization of this framework is discussed by [3].

3 Quasi Maximum Likelihood Estimation of CAW Models

In their paper [10] performs the estimation of model parameters by the method of maximum likelihood. The conditional density of C_t given past information is

$$f(C_t|I_{t-1}) = \frac{|S_t/\nu|^{-\nu/2}|C_t|^{(\nu-n-1)/2}}{2^{\nu n/2}\pi^{n(n-1)/4}\prod_{i=1}^{n}\Gamma\{(\nu+1-i)/2\}} \exp\left\{ -\frac{1}{2}\text{tr}(\nu S_t^{-1}C_t) \right\}.$$

It follows that, up to a constant, the log-likelihood contribution of observation t is given by

$$\ell_t(\theta) = \lambda(v) - \frac{v}{2}\log(|S_t|) - \frac{v}{2}\mathrm{tr}(S_t^{-1}C_t) \tag{8}$$

where

$$\lambda(v) = \frac{vn}{2}\log(v) + \frac{v-n-1}{2}\log(|C_t|) - \frac{vn}{2}\log(2) - \sum_{i=1}^{n}\log[\Gamma\{(v+1-i)/2\}].$$

The last two terms on the right-hand side of (8) are proportional to the value of the Wishart shape parameter v. It immediately follows that the first order conditions for the estimation of a_i and b_j ($i = 1, \ldots, p; j = 1, \ldots, q$), the parameters of the dynamic updating equation for C_t, do not depend on v since the score for observation t with respect to $\theta_c = (a_1, \ldots, a_p, b_1, \ldots, b_q)'$ is given by

$$\frac{\partial \ell_t}{\partial \theta_c} = -\frac{v}{2}\left\{\mathrm{tr}\left(S_t^{-1}\frac{\partial S_t}{\partial \theta_c}\right) - \mathrm{tr}\left(C_t S_t^{-1}\frac{\partial S_t}{\partial \theta_c}S_t^{-1}\right)\right\}. \tag{9}$$

This implies that the value of v is not affecting the estimation of θ_c that can be consistently estimated independently of the value of this parameter. We also note, similar to [3], that, under the assumption that the dynamic model for the conditional expectation of C_t is correctly specified, the score in (9) is a martingale difference sequence even if the Wishart assumption is not satisfied. Analytically, the derivative vector in (9) is given by

$$\frac{\partial \ell_t}{\partial \theta_c} = -\frac{v}{2}\left\{\mathrm{tr}\left(S_t^{-1}\frac{\partial S_t}{\partial \theta_c}\right) - \mathrm{tr}\left(C_t S_t^{-1}\frac{\partial S_t}{\partial \theta_c}S_t^{-1}\right)\right\}. \tag{10}$$

Taking expectations of both sides of (10), conditional on past information I_{t-1}, gives

$$E\left(\frac{\partial \ell_t}{\partial \theta_c}|I_{t-1}\right) = -\frac{v}{2}\left\{\mathrm{tr}\left(S_t^{-1}\frac{\partial S_t}{\partial \theta_c}\right) - \mathrm{tr}\left(E(C_t|I_{t-1})S_t^{-1}\frac{\partial S_t}{\partial \theta_c}S_t^{-1}\right)\right\}. \tag{11}$$

At the true parameter value $\theta_c = \theta_{c,0}$, we have that $E(C_t|I_{t-1}) = S_t$ by application of (2). By substituting this in (11) we obtain

$$E\left(\frac{\partial \ell_t}{\partial \theta_c}|I_{t-1}, \theta_{c,0}\right) = -\frac{v}{2}\left\{\mathrm{tr}\left(S_t^{-1}\frac{\partial S_t}{\partial \theta_c}\right) - \mathrm{tr}\left(S_t^{-1}\frac{\partial S_t}{\partial \theta_c}\right)\right\} = 0.$$

Under the usual regularity conditions, this result allows to interpret $\hat{\theta}_c$, the maximizer, with respect to θ_c, of the log-likelihood obtained from the summation over time of (8), as a Quasi Maximum Likelihood (QML) estimator. Hence, even if the Wishart assumption on the conditional distribution of C_t is not satisfied, the resulting estimator can still be proven to be consistent and asymptotically normal (see [3] for details).

4 Composite Likelihood Estimation of CAW Models

As already discussed in Sect. 1, practical computation of the ML estimator for large dimensional models can be highly computational intensive and, for very large models, even unfeasible. The main reason for this is the necessity of iteratively inverting the high-dimensional covariance matrix S_t in the log-likelihood function at each observation and each iteration of the optimization procedure. Given the values of the time series (T) and cross-sectional (n) dimensions typically considered in multivariate volatility modeling, this operation can be very time consuming. CML estimation offers a practical solution to this problem since the log-likelihood function is approximated by the sum of many low-dimensional marginal log-likelihood functions. In order to face a similar issue arising in the estimation of DCC models, [9] proposes to apply the CML method under the assumption of conditionally normal returns. Differently, in the setting which is being here considered, we derive a CL function for the realized covariance matrix under the assumption of a conditionally Wishart distribution. The derivation of the CL function relies on two results. Before proceeding with their illustration we first need to define the following notations. For any square matrix M_t of order n, we denote by $M_{AA,t}$ a square matrix of order n_A extracted from M_t, which has its main diagonal elements on the main diagonal of M_t. Namely, if A stands for a subset of n_A different indices of $1, 2, \ldots, n$, $M_{AA,t}$ is the matrix that consists of the intersection of the rows and columns of M_t corresponding to the selection of indices denoted by A. The two results can then be formulated as:

R1:   If $C_t \sim W_n(v, S_t/v)$, for any selection of n_A indices we have

$$C_{AA,t} \sim W_{n_A}(v, S_{AA,t}/v).$$

R2:   If $S_t = (1 - \sum_{i=1}^p \alpha_i - \sum_{j=1}^q \beta_j)\bar{C} + \sum_{i=1}^p \alpha_i C_{t-i} + \sum_{j=1}^q \beta_j S_{t-j}$,

$$S_{AA,t} = \left(1 - \sum_{i=1}^p \alpha_i - \sum_{j=1}^q \beta_j\right) \bar{C}_{AA} + \sum_{i=1}^p \alpha_i C_{AA,t-i} + \sum_{j=1}^q \beta_j S_{AA,t-j}.$$

Result 1 is a well known property of the Wishart distribution. By properties of the Wishart distribution, the marginal distribution of $C_{AA,t}$ is also Wishart, with the same degrees of freedom and with scale matrix obtained by deleting from S_t the same rows and columns as in C_t—see [1], Theorem 7.3.4. Applying this result with $n_A = 1$ corresponds to the result that the margin of a diagonal element of a Wishart matrix is a Gamma distribution. Result 2 is an obvious algebraic result.

The CML estimator of the parameters α and β is then defined as the maximizer of the sum of a number of Wishart marginal log-likelihoods for sub-matrices $C_{AA,t}$ corresponding to different choices of indices A. The simplest choice is to select all the log-likelihoods corresponding to sub-matrices of order 2, i.e. to all the $n(n-1)/2$ covariances or pairs of assets. Notice that with these bivariate Wishart log-likelihoods, only matrices of order 2 must be inverted, which can be efficiently programmed. We will denote any Wishart CL functions based on marginal log-likelihoods of dimension 2 as CL_2. Formally, this leads to the following expression

$$CL_2(v, \theta_c, \hat{\hat{C}}) = \sum_{h=2}^{n} \sum_{k<h} \ell_{hk,t}(v, \theta_c, \hat{\hat{C}}_{hk}) \tag{12}$$

with

$$\ell_{hk,t}(.) = \frac{v}{2}\log(v) + \frac{v-3}{2}\log(|C_t^{(hk)}|) - \frac{v}{2}\log(2) - \sum_{i=1}^{2}\log[\Gamma\{(v+1-i)/2\}]$$

$$-\frac{v}{2}\log(|S_t^{(hk)}|) - \frac{v}{2}\mathrm{tr}\{(S_t^{(hk)})^{-1}C_t^{(hk)}\}, \tag{13}$$

where for any matrix M_t, $M_t^{(hk)}$ is the matrix of order 2 extracted at the intersection of rows h and k of M_t. In principle, one can use less terms than the $n(n-1)/2$ terms in (12) without affecting the consistency of the estimator. This can be particularly useful in particular in cases in which n is large. At the price of a slightly more complicated notation, the expression in (13) can be easily generalized to the case in which marginal log-likelihoods of sub-matrices of higher dimension ($m \geq 2$) are used. Under this regard it is important to note that the number of subsystems that can be created, differing for at least one asset, is rapidly increasing with m, making the estimation problem soon unfeasible. The problem is illustrated in Table 1 for different values of m and n. For $n = 50$ assets we have 1,225 bivariate subsystems but 19,600 trivariate subsystems which are different for at least one asset. In the case of $n = 100$ assets, these numbers increase up to 4,950 for $m = 2$ and to 161,700 for $m = 3$. These values further increase for $m > 3$. It follows that for $m > 2$ the Composite Likelihood function can be practically built only using a subsample of all the available marginal likelihoods. Reducing the number of subsystems upon which the composite likelihood is based is, however, expected to reduce efficiency as empirically shown by [9]. On the other hand, increasing the value of m is expected to increase efficiency. For $m = n$, we recover the maximum likelihood estimator as a special case.

Table 1 Number of m—dimensional subsystems versus cross-sectional dimension n

$n \setminus m$	2	3	4	5
10	45	120	210	252
25	300	2,300	12,650	53,130
50	1,225	19,600	230,300	2,118,760
75	2,775	67,525	1,215,450	17,259,390
100	4,950	161,700	3,921,225	75,287,520

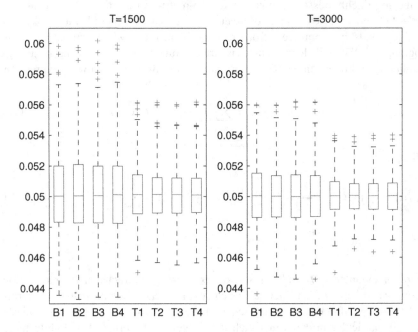

Fig. 1 Simulation results for the estimation of the α parameter by CML with bivariate (B) and trivariate (T) marginals. Key to graph: 1 = estimator based on contiguous sets; 2 = estimator based on $m_1 = [N_m/3]$ subsystems; 3 = estimator based on $m_2 = [N_m/2]$ subsystems; 4 = estimator based on $m_3 = N_m$ subsystems where $N_m = \min(M,5000)$ and $[.]$ indicates rounding to the closest integer

5   Monte Carlo Simulation

In this section we present the results of a Monte Carlo simulation study aimed at evaluating and comparing the finite sample efficiencies of the CML estimator derived under different implementation settings. In particular we will focus on the analysis of CML estimators based on the use of bivariate (CL2) and trivariate marginals (CL3), respectively. The data generating process is assumed to be given by a CAW(1,1) process of dimension $n = 50$ with covariance targeting, as defined by (2) and (7) with parameters given by $\alpha_1 \equiv \alpha = 0.05$, $\beta_1 \equiv \beta = 0.90$ and $\nu = n$. For estimation, $\bar{C}$ is replaced by the sample average $\hat{\bar{C}}$. We generate 500 time series

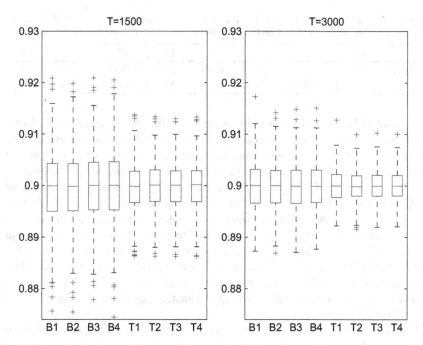

Fig. 2 Simulation results for the estimation of the β parameter by CML with bivariate (B) and trivariate (T) marginals. Key to graph: 1 = estimator based on contiguous sets; 2 = estimator based on $m_1 = [N_m/3]$ subsystems; 3 = estimator based on $m_2 = [N_m/2]$ subsystems; 4 = estimator based on $m_3 = N_m$ subsystems where $N_m = \min(M,5000)$ and [.] indicates rounding to the closest integer

of length $T = 2,000$ and $T = 3,500$, respectively. In both cases the first 500 data points are discarded in order to reduce the impact of initial conditions.

In the bivariate case we compare the estimator based on all the potential bivariate subsystems with three alternatives in which the CL function is built from a subset of the universe of available bivariate systems. First, we consider a subset of dimension $(n - 1)$ composed by the set of contiguous pairs $\{i, i + 1\}$, for $i = 1, \ldots, n - 1$. The rationale behind this choice is that in this way CML estimates are based on the minimal set guaranteeing that all assets are equally represented in the estimation process. Second, we consider two randomly selected subsets of dimension $[M/3]$ and $[M/2]$, respectively, where [.] denotes rounding to the closest integer and M is the overall number of available subsystems. In the trivariate case, considering the whole set of different trivariate systems is not computationally feasible. Hence, the maximum number of trivariate systems which has been considered has been set equal to $N_m = 5,000$. Again, as in the bivariate case, we analyze the sensitivity of the efficiency of the estimator with respect to the number of subsystems used for estimation by considering alternative estimators based on the set of $(n - 2)$ contiguous triplets and on random sets of size $[N_m/3]$ and $[N_m/2]$, respectively. The results have been summarized in Figs. 1, for α, and 2, for β. Both $CL2$ and

Table 2 Simulated variation coefficient (simulated s.e. $(\hat{se})$/simulated mean $(\hat{\mu})$) of parameter estimates for CML estimators based on bivariate (Bi) and trivariate (Ti) marginals, respectively

	B1	B2	B3	B4	T1	T2	T3	T4
$\hat{se}(\hat{a})/\hat{\mu}(\hat{a})$	0.0392	0.0388	0.0388	0.0386	0.0257	0.0244	0.0246	0.0246
$\hat{se}(\hat{b})/\hat{\mu}(\hat{b})$	0.0054	0.0052	0.0052	0.0052	0.0036	0.0033	0.0033	0.0033

$CL3$ result to be approximately unbiased even for the shorter sample size. As far as efficiency of the estimators is concerned, in comparative terms, it is evident that CL3 is remarkably more efficient than CL2 while it appears that, within the range of values considered for the simulation, the number of lower dimensional subsystems used to compute the CL function is not dramatically affecting the efficiency of the CL estimators. In particular, the efficiency gap between CL estimators based on contiguous systems and more complex estimators considering all the systems (bivariate case) or a large random sample of these (trivariate case) appears not substantial. In absolute terms, the efficiency level of both estimators is reasonably high (Table 2). The variation coefficient of CL estimators for parameter α is slightly lower than 4%, for CL2, and approximately equal to 2.5% for CL3 while, for parameter β, the recorded values are approximately equal to 0.5% for CL2 and to 0.3% for CL3.

In this paper only empirical (simulation based) results have been presented. From a theoretical point of view, an issue which has not been addressed is the derivation of the asymptotic distribution of the CL estimators and an estimator of their asymptotic standard errors. Given the complexity of the CAW model, this is not an easy task and we plan to investigate this issue in our future research.

References

1. Anderson, T.W.: An Introduction to Multivariate Statistical Analysis, 2nd edn. Wiley, New York (1984)
2. Bauwens, L., Laurent, S., Rombouts, J.V.K.: Multivariate GARCH models: a survey. J. Appl. Econometrics **21**, 79–109 (2006)
3. Bauwens, L., Storti, G., Violante, F.: Dynamic conditional correlation models for realized covariance matrices, to appear as CORE Discussion Paper (2013)
4. Bonato, M., Caporin, M., Ranaldo, A.: Forecasting realized (co)variances with a block structure Wishart autoregressive model, Working Papers 2009–03, Swiss National Bank (2009)
5. Dovonon, P., Goncalves, S., Meddahi, N.: Bootstrapping Realized Multivariate Volatility Measures. J. Economet. **172**, 49–65 (2013)
6. Engle, R.F.: Dynamic conditional correlation – a simple class of multivariate GARCH models. J. Bus. Econ. Stat. **20**, 339–350 (2002)
7. Engle, R.F.: High dimensional dynamic correlations. In: Castle, J.L., Shephard, N. (eds.) The Methodology and Practice of Econometrics: Papers in Honour of David F Hendry. Oxford University Press, Oxford (2009)
8. Engle, R., Kroner, F.K.: Multivariate simultaneous generalized ARCH. Economet. Theor. **11**, 122–150 (1995)

9. Engle, R.F., Shephard, N., Sheppard, K.: Fitting vast dimensional time-varying covariance models. NYU Working Paper No. FIN-08-009 (2008)
10. Golosnoy, V., Gribisch, B., Liesenfeld, R.: The conditional autoregressive Wishart model for multivariate stock market volatility. J. Eonomet. **167**, 211–223 (2012)
11. Gouriéroux, C., Jasiak, J., Sufana, R.: The Wishart autoregressive process of multivariate stochastic volatility. J. Economet. **150**, 167–181 (2009)
12. Jin, X., Maheu, J.M.: Modelling realized covariances and returns. WP 11-08, The Rimini Center for Economic Analysis (2010)
13. Noureldin, D., Shephard, N., Sheppard, K.: Multivariate High-Frequency-Based Volatility (HEAVY) Models. J. Appl. Economet. **27**(6), 907–933 (2012)
14. Varin, C., Reid, N., Firth, D.: An overview of composite likelihood methods. Statistica Sinica **21**, 5–42 (2011)

A GPU Software Library for Likelihood-Based Inference of Environmental Models with Large Datasets

Michela Cameletti and Francesco Finazzi

Abstract Statistical environmental models are computationally intensive due to the high dimension of the data, both in space and time, and due to the inferential techniques required for parameter estimation and spatial prediction. In particular, the computational complexity of these techniques is related to matrix operations (inversion, solution of linear systems, factorization) involving large dense matrices. Recently, much attention has been paid around the possibility of taking advantage of graphics processing units (GPUs) for mathematical computation. GPUs provide a high degree of parallelism at a reasonable cost and may represent a viable alternative compared to the classic computer cluster configurations. In this work, we develop the shared library GPU4GL implementing ad-hoc linear-algebra functions running on GPUs and compare them with the standard algorithms for CPU. As an example, we apply the GPU functions of GPU4GL to make inference on a non-separable space–time model for air quality data.

1 Introduction

As a consequence of the increasing cultural, social, political and scientific attention paid to environmental problems, there has been a growing availability of data concerning hazardous phenomena for the environment and human health. Environmental phenomena are usually complex and influenced by many factors

M. Cameletti
Department of Mathematics Statistics Computing and Applications, University of Bergamo, Bergamo, Italy
e-mail: michela.cameletti@unibg.it

F. Finazzi (✉)
Department of Information Technology and Mathematical Methods, University of Bergamo, Bergamo, Italy
e-mail: francesco.finazzi@unibg.it

M. Grigoletto et al. (eds.), *Complex Models and Computational Methods in Statistics,*
Contributions to Statistics, DOI 10.1007/978-88-470-2871-5_5,
© Springer-Verlag Italia 2013

that interact with each others and that are linked to climate and anthropogenic pressure. Environmental data, with spatial and/or temporal dimensions, come from heterogeneous sources: monitoring networks, remote sensing, numerical models and geographic information systems. Hence, statistical models should be developed taking into account the complexity of environmental phenomena observed in time and space and characterized by several variability sources. The advantage of the statistical approach consists in taking into account all the possible sources of data and errors and in providing a probabilistic evaluation of the estimate and prediction uncertainty.

In the recent literature on environmental statistical modeling, great emphasis is given to hierarchical models [3, 13], especially in the Bayesian perspective, thanks to their flexibility in decomposing a complicated joint spatio-temporal distribution (which describes the stochastic behavior of the environmental process at all spatial locations and all times, including all possible interactions) into a series of simpler conditional distributions. Spatio-temporal hierarchical models, however, are computationally intensive due to the high dimension of the data in space and time. This is the so-called big n-problem [1, page 387] that is related to the computational costs of linear algebra operations (such as matrix factorization, solution of linear systems, product of matrices) required for model fitting and spatial interpolation. These operations, in fact, involve dense covariance matrices of size n (with n given by the number of observations at all spatial locations and time points) and their computational cost is typically of the order of $\mathcal{O}(n^3)$. Besides, this computational burden gets worse in the Bayesian inference framework where computationally intensive resampling algorithms, such as Monte Carlo Markov chain (MCMC) methods, are employed and linear algebra operations with dense matrices are computed for each iteration of the algorithm.

If we consider that the size of spatio-temporal datasets is generally $n \approx 10^4$ and that the number of iterations required for the MCMC method convergence is of the order of 10^5, the computational complexity of spatio-temporal models and the need for fast computation methods stand out clear. Parallel computing is a viable solution for large statistical problems, such as environmental spatio-temporal hierarchical models with large datasets [8]. In this respect, it is worth noting that MCMC methods do not allow the so-called coarse-grained parallelization as it is necessary to perform sequential updates of the model parameters. Consequently, our interest is focused on fine-grained parallelization of the matrix operations required for each iteration of the algorithm. To achieve the desired results with an acceptable computation time, therefore, it is important to maximize the degree of parallelism. Moreover, since the dense matrices are symmetric and positive definite, efficient algorithms in terms of memory occupation must be adopted.

The potential of parallel processing in statistical computing is well known and the implementation of software and libraries for distributed CPU-based computer clusters is well documented [12, 14]. On the contrary, the graphics processing unit (GPU) technology for parallel computing is relatively recent and its benefits have been explored only for a limited number of statistical applications (see, for example, [9, 11]). The main challenge in using GPU-based algorithms is that they are relatively difficult to program and they require specific programming skills. Hence, we developed the GPU4GL shared library which provides functions for the execution of some linear algebra operations required when making inference with Gaussian likelihoods (GL). Our GPU4GL library can be freely downloaded and loaded in both R and Matlab®, two computing environments widely used in the statistical community.

The aim of this paper is to illustrate the developed GPU4GL library and to investigate the computational benefits related to GPU programming for the implementation of spatio-temporal models for environmental data. In particular, Sect. 2 describes the main features of GPU computing while Sect. 3 is devoted to the description of our GPU4GL library. The computational performances of the library are discussed with reference to a particular matrix operation executed with different hardware and software settings. In Sect. 4 we introduce the case study regarding a spatio-temporal model for particulate matter data in the Piemonte region, Italy. The inference procedures regarding parameter estimation and space–time prediction are discussed in terms of computing time. Conclusions are reported in Sect. 5.

2 GPU Computing

In the development of computer algorithms, the sequential and the parallel design paradigms can be distinguished. The former is easier to understand as it mimics the high-level human reasoning while the latter is often tricky and it usually requires a more complex design phase. The higher complexity of many algorithms, however, is rewarded by a lower computational time.

From a technical point of view, parallel algorithms require parallel computer architectures. For more than two decades, the microprocessor market has been dominated by single-CPU architectures. During that period, software developers and scientists benefited from the increase in the computational speed from one CPU generation to the next. In this sense, the sequential paradigm represented a convenient solution. The parallel paradigm, on the other hand, was relegated to the programming of the few and expensive supercomputers around the world, characterized by hundreds or thousands of CPUs. Around 2003, however, energy consumption and heat-dissipation problems forced the microprocessor vendors to move from the single-CPU architecture to multi-CPU and multi-core architectures. As a consequence, in order to maintain the increasing trend in the computational speed, the algorithm development must embrace the parallel paradigm.

With respect to the past, the main difference is that, nowadays, parallel architectures are widely spread and affordable. It follows that a larger audience can benefit from parallel algorithms and that software developers have incentives to develop parallel code. The multi-CPU/multi-core architecture is more efficient than the old single-CPU architecture. Since the former is a direct evolution of the latter, however, it retains most of the sequential paradigm. In order to fully exploit the advantages of the parallel paradigm, software developers can rely on other parallel architectures.

In this paper, the Compute Unified Device Architecture (CUDA, [7]) developed by Nvidia® is considered. By using the development environment for CUDA, software developers can write parallel code that runs on GPUs of the Nvidia® graphic cards. The GPU is an electronic circuit designed to manipulate and alter the graphic card memory in order to accelerate the building of images in a frame buffer. With respect to a CPU, the GPU is characterized by hundreds of highly-specialized cores, each one performing a small set of simple arithmetic and logic operations. As a consequence, parallel architectures based on GPUs are particularly suitable for sparse and dense matrix algebra operations where few arithmetic operations must be executed over a large number of matrix elements. Note that, though all the operations are executed on the graphic card, the result is transferred to the computer RAM memory rather than being sent to a display.

Dense matrix operations such as matrix multiplication, matrix decomposition and solution of linear systems can be executed with a speedup on the order of 10 to 100x (depending on the algorithm) when compared to homogeneous multi-core architectures. Moreover, as stated by Nvidia®, the CUDA architecture is able to deliver "supercomputing features and performance at 1/10th the cost and 1/20th the power of traditional CPU-only servers." For these reasons, the CUDA architecture is receiving an increasing interest in many scientific fields.

3   The GPU4GL Library

The GPU4GL library is implemented for CUDA and it is based on the MAGMA library v.1.0 (http://icl.cs.utk.edu/magma/), which represents the counterpart for GPU of the LAPACK library. The GPU4GL library is available as Linux compiled shared library at http://code.google.com/p/gpu4gl/ and can be loaded in both R and Matlab® computing environments.

The library implements a set of linear algebra operations commonly used in Gaussian likelihood-based inference techniques. It is worth noting that, though the library is applied here for the analysis of environmental data, its scope is broader. Indeed, Gaussian likelihoods are common in many statistical fields and applications. Moreover, the library applicability is not limited to MCMC methods but it can be extended to other estimation methods such as the EM algorithm when applied to the estimation of space–time hierarchical models (see [4]).

Although the library works even with small matrices, its best computing performances are achieved when the Gaussian likelihood inference involves large matrices. Moreover, the library is a profitable alternative when it is not possible to rely on expensive statistical software environments able to exploit multi-CPU and multi-core architectures. The only drawback of the GPU solution is related to the installed RAM memory on the GPU cards which is often lower than the memory available on off-the-shelf CPU configurations. Indeed, the maximum square matrix dimension manageable by a single GPU ranges from 10,000 to 20,000. However, linear algebra algorithms involving dense square matrices larger than 20,000 are usually unfeasible for both the GPU and the CPU solutions and sparse-matrix or dimension reduction approaches should be considered.

3.1 Library Functions

Let A be a $n \times n$ symmetric and positive definite matrix, B a generic $n \times m$ matrix and $|.|$ the symbol for the matrix determinant. The name, the description of the linear algebra operation and the C++ call[1] of the functions included in the GPU4GL library are the following:

1. The function gpu4gl_cld evaluates the Cholesky decomposition C of the matrix A and computes $l = \log(|A|)$. The C++ function call is

   ```
   gpu4gl_cld(double *A, int *n, double *C, double *l)
   ```

2. The function gpu4gl_lsld evaluates the matrix operation $X = A^{-1}B$ and computes $l = \log(|A|)$. The C++ function call is

   ```
   gpu4gl_lsld(double *A, double *B, int *n, int *m, double *X,
               double *l)
   ```

3. The function gpu4gl_like evaluates the matrix operation $X = B'A^{-1}B$ and computes $l = \log(|A|)$. The C++ function call is

   ```
   gpu4gl_like(double *A, double *B, int *n, int *m, double *X,
               double *l)
   ```

4. The function gpu4gl_rand randomly generates a $n \times 1$ vector **x** from the multivariate Normal distribution $N_n(\mathbf{0}, A)$. The C++ function call is

   ```
   gpu4gl_rand(int *n, double *A, double *x)
   ```

When the library is actually used in a statistical computing environment, it is recommended to implement a wrapper for each function. As an example, the R wrapper for the function gpu4gl_like can be defined as follows:

```
gpu4gl_like <- function(A,B){
    n = nrow(A)
    m = ncol(B)
    res_gpu=.C("gpu4gl_like",A,B,n,m,
        X=as.double(rep(0,n*n)),
        l=as.double(0))
    output = list()
    output$qf = matrix(res_gpu$X,m,m) #quadratic form
    output$logdet = res_gpu$l #log determinant
```

---

[1] Please refer to [6] for more details about the C++ syntax.

```
    return(output)
}
```

where .C() is the R call of the gpu4gl_like library function. The respective R main script should contain the following calls:

```
dyn.load("gpu4gl.so") #load the GPU4GL library
gpu4gl_setdevice(0) #choose the first GPU device
output <- gpu4gl_like(A,B) #call the wrapper function
```

3.2 Computational Performance

In order to compare CPU and GPU performances, the GPU4GL library is tested with respect to several matrix dimensions and four different hardware and software settings. The four considered configurations are:

1. Laptop Intel® Core™ i7 720QM 1.60GHz, 8 GB RAM. Software: Matlab®.
2. Server Dual CPU Intel® Xeon® E5440 2.8GHz, 8GB RAM. Software: Matlab®.
3. Workstation Intel® Xeon® Quad X3430 2.4GHz, 2.5GT/s, 16GB RAM. Software: R.
4. Workstation equipped with NVIDIA® Tesla™ C1060 card 4GB RAM. Software: R and GPU4GL library.

Configuration (1) refers to a standard laptop computer affordable in terms of price. Configuration (2) is related to a more expensive hardware usually available in the computing laboratories of universities and research centers. Configuration (3) is more performant than configuration (1) in terms of computing power but less expensive as it does not include Matlab®. Finally, configuration (4) is based on the CUDA architecture.

With regard to the software configuration, Matlab® exploits the multi-core architecture of the CPUs thanks to the Parallel Computing Toolbox™ while R, by default, uses only one CPU-core.

Table 1 shows the average computing time (in seconds) required for implementing function gpu4gl_like with several matrix dimension values $n = m$. The speedup of configuration (4) with respect to the others is reported in Table 2. Note that 11,680 is the maximum matrix dimension for which the function gpu4gl_like can be executed on the Tesla™ C1060 card. The other matrix dimensions are obtained dividing 11,680 by consecutive powers of 2.

By analyzing the speedup, it can be seen that the GPU configuration is worthy to be used either when large matrices are involved or when the software (R in this case) is unable to exploit the multi-core potentiality offered by the CPU. Note that the speedup decreases when the matrix is large enough to saturate the GPU memory. The speedup for matrices with dimension 11,680 is slightly lower than the speedup for matrices with dimension 5,120.

Table 1 Computing time (in seconds) for running function gpu4gl_like with respect to the four considered system configurations and several matrix dimensions $n \times n$

		Configuration			
		1	2	3	4
Costs in €		2,700	3,000	1,900	1,900
Dimension n	320	0.0044	0.0020	0.0056	0.0032
	640	0.0290	0.0087	0.0414	0.0082
	1,280	0.1702	0.0523	0.3091	0.0381
	2,560	1.1230	0.3288	2.4341	0.2284
	5,120	8.2205	2.3251	19.2193	1.5137
	11,680	100.4612	25.8750	224.7500	19.1460

The cost is the monetary value in € of the hardware and software configuration

Table 2 Speedup for different compared configurations and matrix dimensions $n \times n$

Configuration		Speedup		
		4 vs 1	4 vs 2	4 vs 3
Dimension n	320	1.38	0.62	1.73
	640	3.52	1.06	5.02
	1,280	4.47	1.37	8.13
	2,560	4.92	1.44	10.66
	5,120	5.43	1.54	12.70
	11,680	5.25	1.35	11.74

4 Case Study

The case study concerning air quality of [2] is considered here. The application regards daily data of PM_{10} concentration (particulate matter concentration with an aerodynamic diameter of less than 10 μm in $\mu g/m^3$) collected from October 2005 to March 2006 ($T = 182$ days) by a network of $d = 24$ monitoring stations located in the Piemonte region, Northern Italy. To model the spatio-temporal dynamics of the pollutant concentration we adopt model A3-1 of [2]. The model is characterized by a non-separable covariance function and it includes a large-scale component given by an intercept and by the following covariates: altitude (*alt*, in *m*), coordinates (*utmx* and *utmy*, in km), daily maximum mixing height (*hmix*, in m), daily total precipitation (*rain*, in mm), daily mean wind speed (*ws*, in m/s), daily mean temperature (*temp*, in °K), and daily emission rates of primary aerosols (*emi*, in g/s). The model is characterized by a set of parameters that are estimated in a Bayesian framework through MCMC methods using the GPU4GL library. Once the model has been estimated, it is possible, through predictive posterior distributions, to build daily high-resolution maps of PM_{10} concentration for the Piemonte region.

4.1 The Model

Let $z(s_i, t)$ be the scalar spatio-temporal process observed at site s_i and at time t where $i = 1, \ldots, d$ and $t = 1, \ldots, T$. We assume the following measurement equation

$$z(s_i, t) = u(s_i, t) + \varepsilon(s_i, t) \tag{1}$$

where $\varepsilon(s_i, t) \sim N(0, \sigma_\varepsilon^2)$ is the measurement error defined by a Gaussian white-noise process, serially and spatially uncorrelated. The term $u(s_i, t)$ is the so-called state process given by the sum of a trend and a random process $\omega(s_i, t)$:

$$u(s_i, t) = X(s_i, t) \boldsymbol{\beta} + \omega(s_i, t) \tag{2}$$

where $X(s_i, t) = (X_1(s_i, t), \ldots, X_k(s_i, t))$ denotes the k-dimensional covariate vector for site s_i at time t and $\boldsymbol{\beta} = (\beta_1, \ldots, \beta_k)'$ is the coefficient vector. The zero-mean Gaussian process $\omega(s_i, t)$ is the residual process whose spatio-temporal covariance function depends on the parameter vector θ, namely

$$\text{Cov}\left(\omega(s_i, t), \omega(s_j, t')\right) = \sigma_\omega^2 C_\theta(h, l) \qquad \forall i \neq j, t \neq t' \tag{3}$$

where σ_ω^2 is the constant in time and space variance of the process and $C_\theta(\cdot, \cdot)$ is the spatio-temporal correlation function parameterized by θ, with $h = \|s_i - s_j\|$ the Euclidean distance between sites i and j, and $l = \|t - t'\|$ the temporal lag between time points t and t'. Adopting a non-separable approach as in [5], the space–time correlation function can be defined as:

$$C_\theta(h, l) = \frac{1}{\psi(|l|^2)} \, \varphi\left(\frac{h^2}{\psi(|l|^2)}\right) \tag{4}$$

where $\psi(x) = (ax^\alpha + b)/(b(ax^\alpha + 1))$ and $\varphi(x) = \exp(-cx^\gamma)$, with b, α and γ in $(0, 1]$ and with a and c positive. In this work we consider $a = 0.058$ and $c = 0.549$ fixed and equal to the values estimated in [2].

4.2 Parameter Estimation and Space–Time Prediction

Let $f(.)$ be a probability density function. Moreover, let $\Theta = (\boldsymbol{\beta}, \sigma_\varepsilon^2, \sigma_\omega^2, \alpha, b, \gamma)$ be the parameter vector and Z the collection of all the data $z(s_i, t)$. Using independent prior distributions for the parameters, the log-posterior distribution of Θ given the data is

$$\log\left(f(\Theta \mid Z)\right) \propto -\frac{1}{2}\log|\Sigma| - \frac{1}{2}(Z - X\boldsymbol{\beta})' \, \Sigma^{-1} (Z - X\boldsymbol{\beta}) + \sum_{i=1}^{\dim(\Theta)} \log(f(\Theta_i)) \tag{5}$$

where $X = \{X_1, \ldots, X_T\}$ is the $(dT \times k)$ array of covariates[2] with the t-th element given by $X_t = (X(s_1, t)', \ldots, X(s_d, t)')'$. The (i, j)-th element of the $(dT \times dT)$ dense variance–covariance matrix Σ is defined as

$$\Sigma_{ij} = \sigma_\omega^2 C_\theta \left(\|s_i - s_j\|, \|t_i - t_j\| \right) + \sigma_\varepsilon^2.$$

In order to implement the Gibbs sampling algorithm we derive the full conditional distributions, and when these are not available in an exact closed-form we introduce a Metropolis–Hastings (MH) sampling step (this is the so-called Metropolis-within-Gibbs algorithm, [10]). For more details about the estimation procedure see [2].

From (5) we can easily obtain that the full conditional distribution for β, when a Normal prior distribution $N_k(0, \Sigma_0)$ is assumed, is Gaussian with mean AB' and variance A, where $A = \left(\Sigma_0^{-1} + X' \Sigma^{-1} X \right)^{-1}$ and $B = Z' \Sigma^{-1} X$. To carry out these operations, we use the GPU4GL library as follows:

- function gpu4gl_lsld for computing $\Sigma^{-1} X$, where $A = \Sigma$ and $B = X$;
- function gpu4gl_lsld for computing the inverse of $\left(\Sigma_0^{-1} + X' \Sigma^{-1} X \right)$, where $A = \Sigma_0^{-1} + X' \Sigma^{-1} X$ and B is the identity matrix of size k.

Note that the simulation from $N_k(AB', B)$ can be performed using the standard R function rnorm as k is usually small and, as discussed in Paragraph 3.2, the speedup obtained through the GPU is significant only for large matrices.

All the remaining parameters are estimated through the MH algorithm adopting the function gpu4gl_like with A given by the $(dT \times dT)$ covariance matrix Σ and B the $(dT \times 1)$ residual vector $(Z - X\beta)$.

The spatial prediction of PM_{10} concentration at a new location s_0 and time t_0 (with $1 \leq t_0 \leq T$) is based on the posterior predictive distribution of $z(s_0, t_0)$ which is given by

$$z(s_0, t_0) \mid Z, \Theta \sim N_1 \left(\mathbf{x}(s_0, t_0)\beta + \tilde{\Sigma}' \Sigma^{-1} (Z - X\beta), (\sigma_\omega^2 + \sigma_\varepsilon^2) - \tilde{\Sigma}' \Sigma^{-1} \tilde{\Sigma} \right) \quad (6)$$

where the generic element of the $(d \times 1)$-dimensional vector $\tilde{\Sigma}$ is defined as follows

$$\tilde{\Sigma}_i = \sigma_\omega^2 C_\theta \left(\|s_i - s_0\|, \|t_i - t_0\| \right).$$

Note that the quadratic forms in (6) are also carried out using function gpu4gl_like. When the spatial prediction is jointly performed on a regular grid, the gpu4gl_rand should be considered. In this case, in fact, $\tilde{\Sigma}$ and Σ are large matrices of dimension $d \times m$ and $md \times md$ where m is the grid size.

---

[2]Here braces are used for column stacking of the vectors involved.

Table 3 Posterior estimates (mean and quantities of order 0.025 and 0.975) of the model parameters

	Mean	$q_{0.025}$	$q_{0.975}$		Mean	$q_{0.025}$	$q_{0.975}$
σ_ω^2	0.168	0.154	0.180	β_{hmix}	−0.040	−0.061	−0.019
σ_ε^2	0.029	0.028	0.030	β_{temp}	−0.125	−0.177	−0.075
β_{int}	3.952	3.689	4.211	β_{rain}	−0.060	−0.077	−0.049
β_{alt}	−0.168	−0.238	−0.099	β_{emi}	0.050	0.013	0.088
β_{utmx}	−0.122	−0.200	−0.044	α	0.740	0.681	0.793
β_{utmy}	−0.072	−0.130	−0.013	b	0.184	0.159	0.201
β_{ws}	−0.073	−0.088	−0.059	γ	0.027	0.023	0.032

4.3 Results

In our air quality case study we have that $d = 24$, $T = 182$ and $k = 9$ and the size of the covariance matrix Σ is 4,368. Following the estimation procedure on the previous paragraph, an MCMC simulation of 16,000 iterations has been implemented on R using the GPU4GL library. The first 15,000 iterations have been considered as burn-in and the last 1,000 have been used to compute the posterior mean and the 95% credible interval of each model parameter. The estimation results are reported in Table 3. Note that the parameter values are slightly different from those reported in [2] due to the fact that, in the case of this paper, the parameters a and c are fixed. The average iteration time has been 19.22 s and it is divided as follows: 1.32 s for β, 1.58 s for σ_ε^2, 1.63 s for σ_ω^2, 5.04 s for α and b and 4.61 s for γ. The computing times for α, b, and γ are higher than the computation time of the rest of the parameters. This is due to the fact that a change in the value of the former implies the evaluation of (3) which is implemented by a non-GPU function. When the model estimation is performed on the machine of configuration (2) (server), the average iteration time is 30.94 s, with a speedup equal to 1.61 which is in accordance with the results of Table 2.

As an example, Fig. 1 shows the posterior prediction map for January, 29th 2006 of the PM_{10} concentration evaluated over a regular grid of 56×72 pixels covering the Piemonte region. The map has been obtained considering the posterior mean of all the 1,000 spatial predictions over the grid computed using (6).

5 Conclusions

In this paper, the benefits offered by the GPU technology have been evaluated with respect to the analysis of spatio-temporal environmental data and in particular with respect to the estimation of a statistical spatio-temporal model characterized by a non-separable covariance function. The GPU4GL shared library has been developed and it represents a useful software tool, supporting the GPU technology, for the

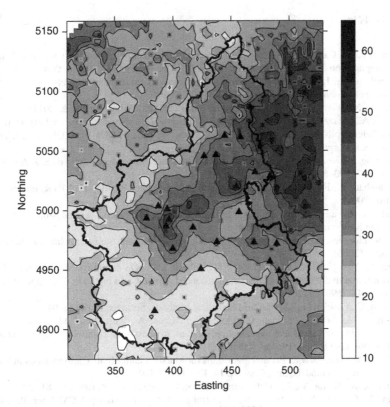

Fig. 1 Particulate matter prediction map for January, 29th 2006

analysis of environmental data and, more generally, for the implementation of Gaussian likelihood-based inference techniques.

The speedup with respect to classic CPU architectures has been assessed to be between around 1.5 and around 12, depending on the number of CPUs and CPU-cores. In this work, the NVIDIA® Tesla™ C1060 card has been considered. The newer (but more expensive) cards, like those of the C20xx family, are seven times faster than the C1060 and they support 6GB memory. Despite the higher price, they represent a valid alternative to the multi-CPU architecture, especially if it is considered that the power consumption of a GPU architecture can be 20 times lower than CPU architectures with the same computational capability. To all the effects, the GPU technology supports the "green computing" as it has a lower impact on the environment when data analysis and simulation, which can last days or weeks, are run.

Acknowledgements This research is part of Project EN17, "Methods for the integration of different renewable energy sources and impact monitoring with satellite data" Lombardy Region—Principal investigator Prof. Alessandro Fassò.

References

1. Banerjee, S., Carlin, B., Gelfand, A.: Hierarchical modeling and analysis for spatial data. Monographs on Statistics and Applied Probability. Chapman and Hall, New York (2004)
2. Cameletti, M., Ignaccolo, R., Bande, S.: Comparing spatio-temporal models for particulate matter in Piemonte. Environmetrics **22**, 985–996 (2011)
3. Cressie, N., Wikle, C.: Statistics for Spatio-Temporal Data. Wiley, New York (2011)
4. Fassò, A., Finazzi, F.: Maximum likelihood estimation of the dynamic coregionalization model with heterotopic data. Environmetrics **22**(6), 735–748 (2011). DOI 10.1002/env.1123. URL http://dx.doi.org/10.1002/env.1123
5. Gneiting, T.: Nonseparable, stationary covariance functions for space–time data. J. Am. Stat. Assoc. **97**, 590–600 (2002)
6. Kernighan, B., Ritchie, D.: The C Programming Language. Prentice Hall PTR, Englewood, Cliffs (2009)
7. Kirk, D., Hwu, W.: Programming Massively Parallel Processors. A Hands-on Approach. Morgan Kaufmann, Los Altos (2010)
8. Kontoghiorghes, E.J. (ed.): Handbook of Parallel Computing and Statistics. Chapman and Hall, London (2006)
9. Lee, A., Yau, C., Giles, M.B., Doucet, A., Holmes, C.C.: On the utility of graphics cards to perform massively parallel simulation of advanced Monte Carlo methods. J. Comput. Graph. Stat. **19**(4), 769–789 (2010)
10. Robert, C., Casella, G.: Monte Carlo Statistical Methods. Springer, Berlin (2004)
11. Suchard, M., Wang, Q., Chan, C., Frelinger, J., Cron, A., West, M.: Understanding GPU programming for statistical computation: studies in massively parallel massive mixtures. J. Comput. Graph. Stat. **19**, 419–438 (2010)
12. Whiley, M., Wilson, S.P.: Parallel algorithms for Markov chain Monte Carlo methods in latent spatial Gaussian models. Stat. Comput. **14**, 171–179 (2004)
13. Wikle, C.K.: Hierarchical models in environmental science. Int. Stat. Rev. **71**, 181–199 (2003)
14. Yan, J., Cowles, M.K., Wang, S., Armstrong, M.P.: Parallelizing MCMC for Bayesian spatiotemporal geostatistical models. Stat. Comput. **17**(4), 323–335 (2007)

Theoretical Regression Trees: A Tool for Multiple Structural-Change Models Analysis

Carmela Cappelli and Francesca Di Iorio

Abstract The analysis of structural-change models is nowadays a popular subject of research both in econometric and statistical literature. The most challenging task is to identify multiple breaks occurring at unknown dates. In case of multiple shifts in mean Cappelli and Reale (Provasi, C. (eds.) S.Co. 2005: Modelli Complessi e Metodi Computazionali Intensivi per la Stima e la Previsione, pp. 479–484. Cleup, Padova, 2005) have proposed a method called ART that employs regression trees to estimate the number and location of breaks. In this paper we focus on regime changes due to breaks in the coefficients of a parametric model and we propose an extension of ART that addresses this topic in the general framework of the linear model with multiple structural changes. The proposed approach considers in the tree growing phase the residuals of parametric models fitted to contiguous subseries obtained by splitting the original series whereas tree pruning together with model selection criteria provides the number of breaks. We present simulation results well as two empirical applications pertaining to the behavior of the proposed approach.

1 Introduction

The detection of regime changes in time series has emerged as an important research topic over the last two decades as proved by several papers published both in the econometric and in the statistics literature (see among the others [3, 4, 8, 14, 17–19]) where the issue of detecting regime changes and the consequences of parameter instability on model specification, inference, and prediction are widely discussed. Indeed, the detection of structural breaks is relevant from several points of view. First it can reveal a behavior of the time series that could otherwise be

C. Cappelli (✉) · F. Di Iorio
TEOMESUS, University Federico II, Naples, Italy
e-mail: carcappe@unina.it; fdiiorio@unina.it

M. Grigoletto et al. (eds.), *Complex Models and Computational Methods in Statistics*,   63
Contributions to Statistics, DOI 10.1007/978-88-470-2871-5_6,
© Springer-Verlag Italia 2013

misunderstood and modeled inadequately, a well-known example is the confusion between long memory and occasional breaks in mean that may lead to an erroneous identification of an integrated or fractionally integrated process (see, e.g., [16, 21]). Second, in the context of forecasting it is sensible to base the forecasts on a model estimated on a recent segment of the series instead of using the entire series especially when we deal with long series covering extended periods of time. Eventually the identification of breaks might isolate shorter periods between longer ones, revealing the presence of outliers and thus the need for adjusting the data (see for example [13]).

Since the seminal paper of [1] that addressed the case of a single break occurring at an a priori unknown break date, the focus is on detecting multiple breaks at unknown dates that represents the most challenging task. In this context the undiscussed contribution is due to Bai and Perron that in various papers [5–7] have presented a comprehensive discussion of the issue providing estimation methods, testing procedures and confidence intervals for multiple structural changes in the framework of the linear model.

In case of level shifts [12] has proposed a computationally efficient procedure called Atheoretical Regression Trees (ART) to detect multiple breaks in mean occurring at unknown dates that is based on Least Square Regression Trees (so forth denoted LSRT) and it mimics Bai and Perron's estimation method of the break dates.

Although ART is a completely heuristic procedure and thus it is not possible to conduct inference on the break dates, extensive simulation studies and comparison with current methods have provided evidence of its usefulness. The results as well as applications to various real-time series can be found in [20] whereas in [11] ART is employed in an empirical procedure that helps to distinguish between long memory and breaks in the mean whose large memory and computing requirements benefit from a fast approach such as ART to locate the breaks.

Up to now ART has been used to detect shifts in mean whereas in this paper the focus is on parameter instability over time, i.e. the case when the coefficients of a parametric model are subject to change considering the general framework of the linear model with multiple structural changes.

At this aim we propose a novel version of the ART procedure that we call Theoretical Regression Trees (TRT) because it extends the principle of recursive partitioning for dating the breaks to parametric models fitted to contiguous segments obtained by splitting the data. Then, tree pruning together with model selection criteria provides the actual number of breaks.

The remainder of the paper is organized as follows. In Sect. 2 we discuss the issue of estimating multiple breaks from a general point of view and we introduce the ART method showing how it can be employed to detect multiple changes in mean. Then we illustrate how ART can be extended giving raise to TRT that accounts for parameter instability in the linear model. In Sect. 3 we report the results of various simulation experiments pertaining to the behavior of the procedure whereas in Sect. 4 we present two empirical applications that illustrate its practical usefulness. Final remarks follow in Sect. 5.

2 Estimation of Multiple Structural Changes with Theoretical Regression Trees

2.1 Background

The issue of estimating multiple structural breaks can be briefly illustrated as follows. Let y_t be a time series characterized by $m + 1$ regimes and m breaks so that $t = T_{j-1} + 1, \ldots, T_j$ and $j = 1, \ldots, m + 1$ (we adopt the common convention that $T_0 = 0$ and $T_{m+1} = T$ where T is the length of the series). A common estimation method of the set of unknown break dates is that based on the least square principle, i.e., the estimated break points $(\hat{T}_1, \ldots, \hat{T}_m)$ are such that:

$$(\hat{T}_1, \ldots, \hat{T}_m) = \text{argmin}_{(T_1, \ldots, T_m)} \text{SSR}(T_1, \ldots, T_m) \tag{1}$$

where $\text{SSR}(T_1, \ldots, T_m)$ denotes the sum of squared residuals of the partition that in case of multiple shifts in mean is given by:

$$\text{SSR}(T_1, \ldots, T_m) = \sum_{j=1}^{m+1} \sum_{t=T_{j-1}+1}^{T_j} (y_t - \mu_j)^2. \tag{2}$$

To detect the presence of such structural changes [12] has proposed a procedure based on LSRT [9] which are piecewise-constant models: a node h is split into its left and right descendants h_l and h_r to reduce the deviance of the dependent variable y fitting to each node the mean of corresponding y's values. The algorithm selects the split, i.e. the binary division, that minimizes:

$$\text{SSR}(h_l) + \text{SSR}(h_r) = \sum_{g \in \{l,r\}} \sum_{y \in h_i} (y_t - \hat{\mu}(h_g))^2 \tag{3}$$

where $\hat{\mu}(h_g)$ is the mean of the y values in node h_g ($g \in \{l, r\}$) thus, the splitting criterion (3) corresponds to the objective function (2) computed for a binary partition. Figure 1 displays the procedure for a single split in a binary tree diagram. Once the partition of a node is performed, the splitting process is recursively applied to each subnode until either the subnodes reach a minimum size or no improvement of the criterion can be achieved.

As shown by [12] LSRT provide a practical tool for dating multiple shifts in mean occurring at unknown dates. Indeed, given an observed time series y_t whose breaks in mean we want to identify, by tree regressing y_t on a sequence of completely ordered numbers $i = 1, \ldots, T$ we obtain a partition of the series into contiguous segments such that $\hat{\mu}_j \neq \hat{\mu}_{j+1}$; the partition is represented as a binary tree whose split points identify candidate break dates whereas the terminal nodes of the tree provide the regimes. The procedure called ART mimics the well-known break dates estimation method proposed by [5, 6], providing comparable results while being

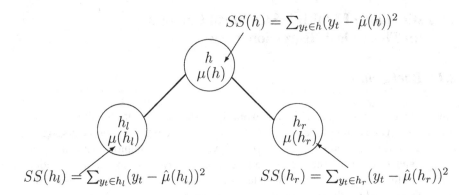

$$SS(h) = \sum_{y_t \in h}(y_t - \hat{\mu}(h))^2$$

h
$\mu(h)$

h_l
$\mu(h_l)$

h_r
$\mu(h_r)$

$$SS(h_l) = \sum_{y_t \in h_l}(y_t - \hat{\mu}(h_l))^2 \qquad SS(h_r) = \sum_{y_t \in h_r}(y_t - \hat{\mu}(h_r))^2$$

Fig. 1 Tree diagram

much faster because it is a local minimizer of the objective function (for discussion on this issue see [20]).

2.2 Theoretical Regression Trees

Let us consider the following linear regression model with multiple structural-changes:

$$y_t = x_t'\beta_j + \epsilon_t \quad (t = T_{j-1} + 1, \ldots, T_j, \; j = 1, \ldots, m + 1)$$

where y_t is the observed dependent variable at time t, x_t is a $(k \times 1)$ vector of regressors, β_j is the $(k \times 1)$ vector of regression coefficients of the j-th regime and ϵ_t is the disturbance at time t. The goal is to estimate $(\beta_1, \ldots, \beta_{m+1}, T_1, \ldots, T_m)$, i.e. both the unknown break dates and coefficients.

In order to achieve this goal the binary recursive partitioning approach of ART can be employed. In this case at each tree node h the best binary split is selected considering the residuals of proper regression models fitted to contiguous segments, thus two separate regressions are estimated for each value ($\min_{\text{obs}} \le i \le T(h) - \min_{\text{obs}}$) where $T(h)$ is the length of subseries in node h and $\min_{\text{obs}}$ denotes a minimum number of observations required to estimate the model. The best split of node h minimizes the sum of squared residuals

$$SSR(h_l) + SSR(h_r) = \sum_{g \in \{l,r\}} \sum_{y_t \in h_g} (y_t - x_t'\hat{\beta}(h_g))^2$$

where $\hat{\beta}(h_g)$ is the estimate of the regression coefficients at subnode h_g ($g \in \{l, r\}$). The selected split point is the estimated break date of the subseries in node h and the splitting process, recursively applied to each subseries obtained by cutting off at the estimated break date, produces a large binary tree. Then, cost-complexity pruning [9] is employed to generate a sequence of subtree, i.e. of nested partitions that are alternative structural-change models of various dimension (number of breaks and regimes). A common procedure to select the preferable subtree (model) among the competing ones is to consider an information criterion; in particular we use the Bayesian Information Criterion (BIC) defined as:

$$\text{BIC}(m) = -2\text{log}lik + p\text{log}(T)$$

where $p = (k + 1) \times (m + 1)$ (for the computation of other common information criteria in regression trees, see [22]).

We call the proposed approach TRT because, opposed to ART that detects level shifts without modeling the data, in the present case the break points are identified estimating a parametric model in the segments.

3 Simulation Experiments

In this section we present the results of simulation experiments carried out to evaluate the performance of TRT in terms of either number of structural changes detected (nb) or rate of correct identification of the break dates (ci) considering the exact identifications as well as short intervals around the true value. In the spirit of [7] various data generating processes are considered allowing for different types of changes in intercept and/or in slope as well as serial correlation.

We start with the case where no serial correlation is present, the basic data generating processes used are:

DGP-1: $y_t = \epsilon_t$;

$$\text{DGP-2:} \begin{cases} y_t = \mu_1 + \beta_1 x_t + \epsilon_t & \text{if} \quad t \leq 75; \\ y_t = \mu_2 + \beta_2 x_t + \epsilon_t & \text{if} \quad t > 75; \\ y_t = \mu_3 + \beta_3 x_t + \epsilon_t & \text{if} \quad t > 90. \end{cases}$$

Throughout $\{\epsilon_t\}$ denotes a sequence of $i.i.d.$ $N(0, 1)$ random variables, $\{x_t\}$ a sequence of $i.i.d.$ $N(1, 1)$ random variables uncorrelated with $\{\epsilon_t\}$ and 1,000 Monte Carlo replications are generated. Note that DGP-1 exhibits no structural changes and thus it is the base case to assess the ability of the method to select the correct number of breaks. For DGP-2 and DGP-3 alternative parameter values are considered that are reported in Table 1 where the corresponding results are presented.

The results for DGP-1 where the series is white noise show that when no breaks are present in the data the method occasionally detects spurious breaks but increasing the sample size reduces such erroneous identifications. For DGP-2, the case with one break, the BIC computed on the sequence of pruned subtrees,

Table 1 Simulation results

	$m = 0, 1$				
	nb	ci	$ci \pm 1$	$ci \pm 2$	$ci \pm 4$
DGP-1					
$T = 125$	0.05	–	–	–	–
$T = 250$	0.03	–	–	–	–
DGP-2 ($T = 150$)					
$\mu_1 = 1, \mu_2 = 2.0$ $\beta_1 = 1.4, \beta_2 = 1.6$	1.00	0.30	0.45	0.64	0.77
$\mu_1 = 1, \mu_2 = 2.5$ $\beta_1 = 1.4, \beta_2 = 1.8$	1.00	0.52	0.74	0.85	0.93

		$m = 2$							
		Break 1 ($T_2 = 90$)				Break 2 ($T_1 = 45$)			
	nb	ci	$ci \pm 1$	$ci \pm 2$	$ci \pm 4$	ci	$ci \pm 1$	$ci \pm 2$	$ci \pm 4$
DGP-3 ($T = 150$)									
$\mu_1 = 1, \mu_2 = 2, \mu_3 = 0.5$ $\beta_1 = 1, \beta_2 = 1.5, \beta_3 = 0.5$	1.81	0.39	0.63	0.78	0.89	0.20	0.47	0.59	0.72
$\mu_1 = 1, \mu_2 = 2, \mu_3 = -0.5$ $\beta_1 = \beta_2 = 0.5, \beta_3 = 1$	1.88	0.60	0.91	0.95	0.98	0.24	0.51	0.66	0.76
$\mu_1 = 1, \mu_2 = 2.5, \mu_3 = -0.5$ $\beta_1 = \beta_3 = 1, \beta_2 = 0.5$	2.00	0.81	0.98	1.00	–	0.49	0.67	0.76	0.91

chooses a single break 100% of the times for both settings. As to the rate of correct identifications, it is adequate for the first setting where the shifts, especially in slope, are quite mild, and it tends to become higher as the size of the break increases, as we can see from the second setting where a relatively slight increase of the shifts causes the percentage of correct identifications to reach 52%, 74% within 1 observation, and 93% within four observations. A similar picture emerges for DGP-3 where two breaks are present. Here we have considered three alternative settings characterized by a somewhat increasing break sizes. In all cases the break at time $T_2 = 90$ is the strongest and the first to be identified (thus denoted break 1 in Table 1). Starting with the mean number of breaks detected we can see that in the first two cases the method slightly underestimate the true number whereas for the third setting the number is correct. Coming to the rate of correct identifications of the break dates, for the strongest break ($T_2 = 90$) the percentages are in all cases pretty high, but, as expected, lower for the second break identified ($T_1 = 45$) especially for the first two settings but it becomes higher in the third setting where the break is larger.

These results show the proposed approach to perform quite well in terms of right number of breaks detected and the rate of correct identifications of the break date and also that, although the sample size is relatively small, the precision of the method improves considerably when adequate sized breaks are considered.

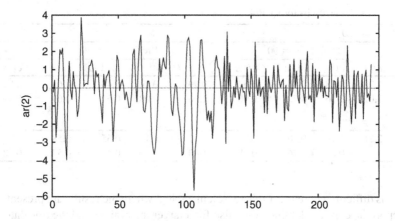

Fig. 2 An illustrative simulated series

In a further simulation experiment we have considered the case when serial correlation is present and the breaks are not equally spaced. Also, for the purpose of making the simulation study more insightful, we have compared TRT with the estimation method of the break dates proposed by Bai and Perron that is a global minimizers of the sum of squared residuals.

We generated 1,000 series from the following DGP:

$$\text{DGP-4:} \begin{cases} y_t = \phi_{11} y_{t-1} + \phi_{21} y_{t-2} + \epsilon_t & \text{if} \quad t \le T_1; \\ y_t = \phi_{12} y_{t-1} + \phi_{22} y_{t-2} + \epsilon_t & \text{if} \quad T_1 + 1 < t \le T_2; \\ y_t = \phi_{13} y_{t-1} + \phi_{23} y_{t-2} + \epsilon_t & \text{if} \quad T_2 + 1 < t \le T; \end{cases}$$

where ϵ_t denotes a sequence of $i.i.d.$ $N(0, 1)$ random variables. DGP-4 which is an AR(2) with two breaks is a case where serial correlation is taken into account parametrically.

We have used sample sizes of $T = 120$ and $T = 240$. The break dates are $T_1 = 30$ and $T_2 = 90$ for sample size $T = 120$ and $T_1 = 75$ and $T_2 = 125$ for sample size $T = 240$. The parameter values in the three regimes are: $\phi_{11} = 0.6$ $\phi_{12} = 1.2$ $\phi_{13} = -0.4$ and $\phi_{21} = -0.6$ $\phi_{22} = -0.6$ $\phi_{23} = -0.2$. Note that in DGP4 no constants are included in the different regimes because, in general, the presence of changes in the constant tends to favor the identification of breaks whereas our focus is on the ability of the method to detect breaks in the regressors coefficients.

For illustrative purposes in Fig. 2 one of the $AR(2)$ simulated series is depicted. As it can be seen the graphical inspection of series suggests the presence of a single break that corresponds to the third regime when both autoregressive parameters turn to negative. In particular, although the second regime is characterized by a more marked sinusoidal behavior, it is not distinguishable from the first one. Indeed, we have set up on purpose a case where the detection of the number and location of

Table 2 Simulation results for DGP-4

| | $T = 120$ | | | | | | | |
| | Break 1—$t = 90$ | | | | Break 2—$t = 30$ | | | |
	nb	ci	$ci \pm 2$	$ci \pm 4$	$ci \pm 8$	ci	$ci \pm 2$	$ci \pm 4$	$ci \pm 8$
TRT	1.93	0.18	0.83	0.94	0.99	0.10	0.40	0.61	0.81
BP	1.01	0.16	0.72	0.90	1.00	0.08	0.54	0.69	0.80
	$T = 240$								
TRT	2.21	0.19	0.70	0.90	0.97	0.12	0.35	0.58	0.77
BP	1.04	0.03	0.76	0.88	1.00	0.10	0.44	0.57	0.78

breaks is difficult whereas for this reason in Table 2 where the results are presented, we report the correct identifications also for a larger interval around the true date.

As a general observation both methods show very low rates of exact identification of the break date even for the strongest break (break 1 at date T_2). This finding seems to be due to the presence of autocorrelation; however, TRT slightly outperforms BP. Indeed, as a local optimizer, TRT estimates the breaks one at time and it generates nested partitions thus, once a break has been detected its date cannot be changed. In our experience this feature makes TRT more stable and less sensitive to the data with respect to the global optimizers of BP.

Considering the various intervals around the true dates, the rates of correct identification associated with both methods are high for the larger break at date T_2 and adequate for the smaller one occurring at date T_1. Occasionally one method outperforms the other but on the whole the simulation results confirm that our approach provides comparable results. Another interesting feature is that doubling the sample size does not lead to more accurate estimates of the break dates, suggesting that what is important is the size of the break rather than the sample size.

As to the number of breaks identified, it is worth noticing that, it depends on the selection criterion but, although both methods employ the BIC, the number of breaks is well identified by TRT but, surprisingly, it is underestimated by BP. Indeed, in various simulation studies Bai and Perron report the BIC to choose a number of breaks much higher than the true value. This issue deserves further investigation as well as the use of alternative selection criteria.

4 Empirical Applications

In this section we present two empirical applications of TRT. The first analyzes one of the series considered in the study of [10] on the existence of a long-run relationship between investment and savings in a panel of 18 OECD countries. The second reevaluates the findings of [23] related to the detection of changes in the time series of the car drivers killed or seriously injured in Great Britain in traffic accidents.

4.1 Structural Changes in Savings–Investment Relationship: The Case of Finland

According to the fundamental macroeconomic relationship Gross Domestic Product (GDP) Y, consumption C, capital formation I and current account B are linked by the basic relationship $Y = C + I + B$. Moreover it is well known that in an open economy capital formation is not constrained by domestic savings. Nevertheless [15] documented a close correlation between savings and investments ratios to GDP in 16 OECD economies; a way to model this evidence, called "Feldstein-Horioka puzzle" (for discussion and details on the topic, see [2]), is given by the FH equation

$$i_t = \mu + \beta s_t + \epsilon_t$$

where $i = \log(I/\text{GDP})$ and $s = \log(S/\text{GDP})$. Since in closed economies investments are, by definition, equal to savings, the coefficient β, known as "saving-retention rate," is constrained to be equal to 1, the constant μ to 0, and the residuals, that reflect errors of measurement, are stationary.

Actually, the existence of a strict relationship between investment and saving can be assumed when capital movements are strictly regulated but, over the last few decades the regulations underwent many changes and, in particular, within the European Union controls were removed in 1990. Hence, a more plausible model that allows for a change in the coefficients is defined as (see e.g. [10])

$$i_t = \begin{cases} \mu_0 + \beta_0 s_t + \epsilon_t, & t \le t^* \\ \mu_1 + \beta_1 s_t + \epsilon_t, & t > t^*. \end{cases} \tag{4}$$

We have employed model 4 to analyze the case of savings–investment relationship in Finland treating the break date as unknown because, although the regulation changes suggest possible break dates, in general, these cannot be assumed as a priori known due to delay effects. Thus, we have employed TRT to estimate the unknown break date t^*.

Figure 3 plots the dataset that is freely available from the OECD.stat database (see "National Accounts," table "Disposable income and net lending—net borrowing"). The series considered are: Investment (as Gross capital formation—transaction code P5S1) and Savings (obtained as sum of Net savings—transaction code: B8NS1—and Consumption of Fixed capital—transaction code K1S1) as log-ratio to GDP (transaction code B1-GS1), in national currency at current prices. The sample period is 1970–2011.

As we can see, the two variables follow a closely related path up to the early 1990s when the association weakens till the 2007–2008 (when the financial crises started) but a possible second break so close to the end of the series would be not trustful.

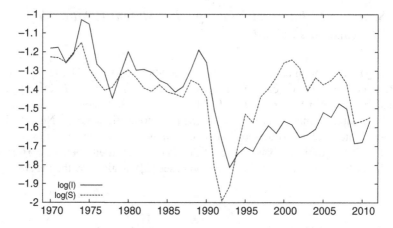

Fig. 3 Finland: Savings/GDP (*dashed line*) and Investment/GDP (*solid line*), 1970–2011

Table 3 Estimates of the model on the entire series and on the regimes identified by the breaks

Entire series	β_0	β_1
1970–2011	−0.466**	0.693***
Regimes		
1970–1992	−0.392***	0.651***
1993–2011	−1.051***	0.397***

Significance codes: ∗∗∗ = 0.001, ∗∗ = 0.01, ∗ = 0.05, · = 0.1

Indeed TRT founds evidence of a break in 1993 (note that the same break date is identified with Bai and Perron procedure). The OLS estimates[1] associated with the two sub-samples are reported in Table 3.

We see that the estimates vary across the regimes confirming the graphical evidence, in particular the saving-retention rate decreases by 40% shifting from the first to the second regime.

4.2 The UK Seatbelt Data

The issue of interest is the effect of making compulsory the use of seat belts by vehicle occupants. The data are monthly and the sample is 1969:1-1984:12 ($T =$ 192). The graphical inspection of the series, depicted in Fig. 4 suggests, the presence of two breaks.

---

[1]Note that estimation methods such as FM-OLS that allow for nonstationarity have not been considered due to the short sample size in the two sub-regimes.

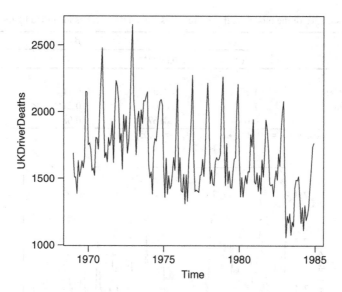

Fig. 4 The Seatbelt data

Reference [23] analyzed this series in a least square framework considering the logarithm of the data and regressing it on its lagged values at lag 1 and 12 using ordinary least squares. The model is:

$$\log(y_t) = \phi_0 + \phi_1 \log(y_{t-1}) + \phi_2 \log(y_{t-12}) + \epsilon_t.$$

By employing structural change tests the authors found evidence of at least one break and they favored a model with $m = 2$ breaks at dates 1973:10 and 1983:1. The first break can be associated with oil crisis and the second, indeed, with the introduction of compulsory seatbelt wearing in UK. Bai and Perron estimation procedure based on global optimizers identifies the same break dates but the BIC chooses the model with no breaks.

We have performed change point analysis of the series with TRT considering the same regression model that involves the lagged regressors adding also a trend. TRT detects the same break dates at 1973:10 and 1983:1 but for our model the corresponding partition provides the minimum BIC. It is worth noticing that the first (strongest) break to be detected is 1983:1 and that the values of the BIC computed for $m = 1, 2$ are very close to each other suggesting that either models are plausible for the data.

The coefficient estimates computed from the whole series and the three regimes identified by the break points are reported in Table 4 whereas Fig. 5 depicts the log-transformed data and the fitted response variable for the model with $m = 2$ breaks.

Table 4 Estimates of the model on the entire series and on the regimes identified by the breaks

Entire series	β_0	β_1	β_2	β_3
1969:1-1984:12	1.06*	−0.0004**	0.37***	0.49***
Regimes				
1969:12-1973:10	1.57˙	0.0003	0.11	0.68***
1973:11-1983:1	1.18*	0.0006*	0.21**	0.62***
1983:2-1984:12	−0.04	0.0167***	0.08	0.53***

significance codes: $\star\star\star = 0.001, \star\star = 0.01, \star = 0.05, \cdot = 0.1$

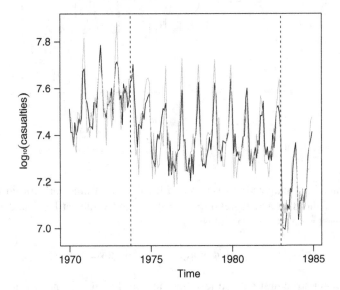

Fig. 5 The log-transformed seatbelt time series, the fitted values, and the break dates

We see that the estimates of the coefficients vary across the three regimes and that the fitting provided by the model seems quite good.

5 Conclusions

This paper has addressed the issue of detecting multiple structural changes in the general framework of the liner model presenting a method, TRT, that employs the recursive partitioning approach of regression trees. The method allows to estimate the unknown number and location of change points by fitting parametric models to contiguous segments arising by splitting the data.

For the purpose of evaluating the performance of the proposed approach we have carried out simulation experiments. According to the results TRT selects the correct number of breaks accurately and its precision location especially when the break size is adequate, and thus the method can be confidently used to analyze structural-change models.

As a further benchmark we have applied our procedure to the regime change analysis of real-time series: the UK seat-belt data and the savings–investment data of Finland. In both cases the results have shown that our procedure is able to reveal relevant features of the series and thus it represents a useful tool to analyze models with multiple breaks.

The procedure can be easily implemented in any software that provides the classification and regression tree methodology and it provides a quick flexible tool, that, due to its simplicity, is particularly useful for applied time series analysis.

Further research will address the use of alternative criteria to select the actual number of breaks as well as issues involving nonstationarity and nonlinearity.

Acknowledgments Paper partially supported by MIUR grant (code 2008WKHJPK-PRIN2008-PUC number E61J10000020001)

References

1. Andrews, D.W.K.: Tests for parameter instability ans structural change with unknown change point. Econometrica **61**, 821–856 (1993)
2. Apergis, N., Tsoumas, C.: A survey of the Feldstein Horioka puzzle: what has been done and where we stand. Res. Econ **63**, 64–76 (2009)
3. Bai, J.: Least square estimates of a shift in linear processes. J. Times Ser. Anal. **15**, 453–472 (1994)
4. Bai, J.: Estimating multiple breaks one at time. Economet. Theor. **15**, 315–352 (1997)
5. Bai, J., Perron, P.: Estimating and testing linear models with multiple structural changes. Econometrica **66**, 47–78 (1998)
6. Bai, J., Perron, P.: Computation and analysis of multiple structural change models. J. Appl. Economet. **18**, 1–22 (2003)
7. Bai, J., Perron, P.: Multiple structural change models: a simulation analysis. In: Broy, M., Dener, E. (eds.) Econometric Theory and Practice: Frontiers of Analysis and Applied Research, pp. 212–237. Cambridge University Press, Cambridge (2006)
8. Banerjee, A., Urga, G.: Modelling structural breaks, long memory and stock market volatility: an overview. J. Economet. **129**, 1–34 (2005)
9. Breiman, L., Friedman, J.H., Olshen, R.A., Stone, C.J.: Classification and Regression Trees. Wadsworth & Brooks, Monterey (CA) (1984)
10. Di Iorio, F., Fachin, S.: Savings and Investments in the OECD, 1970–2007: a Panel Cointegration test with breaks. MPRA Paper 3139, University Library of Munich, Germany (2010)
11. Cappelli, C., Di Iorio, F.: Structural breaks versus long memory: a simulation study. Stat. Appl. **17**, 285–295 (2007)
12. Cappelli, C., Reale, M.: Dating multiple structural breaks occurring at unknown dates via Atheoretical Regression Trees. In: Provasi, C. (eds.) S.Co. 2005: Modelli Complessi e Metodi Computazionali Intensivi per la Stima e la Previsione, pp. 479–484. Cleup, Padova (2005)
13. Cappelli, C., Penny, R.N., Rea, W., Reale, M.: Detecting multiple mean breaks at unknown points with regression trees. Math. Comput. Simul. **78**, 351–356 (2008)
14. Chong, T.: Structural change in AR(1) models. Economet. Theor. **17**, 87–155 (2001)
15. Feldstein, M., Horioka, C.: Domestic saving and international capital flows. Econ. J. **90**, 314–329 (1980)

16. Granger, C.W.J., Hyung, J.: Occasional structural breaks and long memory with an application to the S&P 500 absolute stock returns. J. Emp. Fin. **11**, 399–421 (2004)
17. Hansen, B.: The new econometrics of structural change: dating breaks in U.S. labor productivity. J. Econ. Persp. **15**, 117–128 (2001)
18. Hansen, B.: Testing for parameter instability in linear models. J. Policy Model. **14**, 517–533 (1992)
19. Pesaran, M.H., Pettenuzzo, D., Timmermann, A.G.: Forecasting time series subject to multiple structural breaks. Rev. Econ. Stud. **73**, 1057–1084 (2006)
20. Rea, W., Reale, M., Cappelli, C., Brown, J.A.: Identification of changes in mean with regression trees: an application to market research. Economet. Rev. **29**, 754–777 (2010)
21. Smith, A.: Level shift and the illusion of long memory in economics time series. J. Bus. Econ. Stat. **23**, 321–335 (2005)
22. Su, X.G., Wang, M., Fan, J.J.: Maximum likelihood regression trees. J. Comput. Graph. Stat. **13**, 586–598 (2004)
23. Zeiles, A., Kleiber, C., Kramer, W., Hornik, K.: Testing and dating of structural changes in practice. Comput. Stat. Data Anal. **44**, 109–123 (2003)

Some Contributions to the Theory of Conditional Gibbs Partitions

Annalisa Cerquetti

Abstract Conditional Gibbs partitions naturally arise in Bayesian nonparametric analysis of species sampling problems under almost surely discrete priors inducing infinite exchangeable partitions with distribution in Gibbs product form. In this setting interest relies on posterior predictive inference on some characteristics of a population of species, given an initial sample of observations. Here we focus on the subclass of Poisson-Kingman partitions driven by the Stable subordinator, and, relying on the unconditional theory of exchangeable Gibbs partitions, derive some additional results for the posterior partition, the conditional α diversity and a Stirling's approximation of the Gibbs weights.

1 Introduction

Exchangeable Gibbs partitions [13] are a class of infinite exchangeable partitions of the positive integers characterized by having distribution, expressed by a consistent sequence of symmetric functions p of compositions of n, called *exchangeable partition probability function (EPPF)*, in the following Gibbs product form

$$p(n_1, \ldots, n_k) = V_{n,k} \prod_{j=1}^{k} (1 - \alpha)_{n_j - 1}, \tag{1}$$

for each $n \geq 1$, $k \leq n$, $\sum_{j=1}^{k} n_j = n$, $\alpha \in (-\infty, 1)$, and the $(V_{n,k})$ weights satisfying the backward recursive relation $V_{n,k} = (n - k\alpha)V_{n+1,k} + V_{n+1,k+1}$ with $V_{1,1} = 1$. Here $(x)_y$ stands for rising factorials: $(x)_y = (x)(x+1)\cdots(x+y-1) = \Gamma(x+y)/\Gamma(x)$. The fundamental result in [13] (Theorem 12) states that each

A. Cerquetti (✉)
MEMOTEF, Sapienza University of Rome, Rome, Italy
e-mail: annalisa.cerquetti@uniroma1.it

M. Grigoletto et al. (eds.), *Complex Models and Computational Methods in Statistics*,
Contributions to Statistics, DOI 10.1007/978-88-470-2871-5_7,
© Springer-Verlag Italia 2013

element in the exchangeable Gibbs class arises as a mixture of corresponding extreme partitions being, respectively: *Fisher's partitions* [11] for $\alpha < 0$, *Ewens' partitions* [6] for $\alpha = 0$ and conditional *Poisson–Kingman partitions driven by the stable subordinator* [26] for $\alpha \in (0, 1)$. By *Kingman's correspondence* [17]

$$p(n_1, \ldots, n_k) = \sum_{(j_1, \ldots, j_k)} \mathbb{E}\left[\prod_{i=1}^{k} P_{j_i}^{n_1}\right]$$

where $(j_1, \ldots, j_k)$ ranges over all ordered k-tuples of distinct positive integers, the Gibbs characterization of Gnedin and Pitman still holds if stated in terms of the random atoms in the infinite sum representation of the corresponding random probability measure $P(\cdot) = \sum_{j=1}^{\infty} P_j \delta_{X_j}(\cdot)$, for X_j iid $\sim H(\cdot)$ nonatomic, independent of the $(P_j)_{j \geq 1}$ taking values in the infinite dimensional simplex. See [27] for a comprehensive reference on those topics and related combinatorial processes.

Here we focus on exchangeable Gibbs partitions induced by sampling from a random probability measure whose *ranked* atoms $(P_j^{\downarrow})$ follow a Poisson–Kingman distribution $\text{PK}(\rho_\alpha, \gamma)$ driven by the Lévy density of the stable subordinator $\rho_\alpha(x) = \alpha \Gamma(1 - \alpha)^{-1} x^{-\alpha-1} dx$, for $0 < \alpha < 1$ and $x > 0$ ([26]). Here γ stands for some mixing density on $(0, \infty)$ that, without loss of generality, may be expressed as $h(t) f_\alpha(t)$ for $f_\alpha(\cdot)$ the stable density, so that

$$\text{PK}(\rho_\alpha, h \times f_\alpha) = \int_0^{\infty} \text{PK}(\rho_\alpha | t) h(t) f_\alpha(t) dt$$

where $\text{PK}(\rho_\alpha | t)$ is the regular conditional distribution of $(P_j^{\downarrow})$ given $T = t$. Equivalently, in terms of EPPFs, each $\text{PK}(\rho_\alpha, h \times f_\alpha)$ corresponds to

$$p_{\alpha,h}(n_1, \ldots, n_k) = \int_0^{\infty} p_\alpha(n_1, \ldots, n_k | t) h(t) f_\alpha(t) dt \tag{2}$$

where (cfr. [26], (66)), for $s = t^{-\alpha}$

$$p_\alpha(n_1, \ldots, n_k | s^{-1/\alpha}) = \frac{\alpha^k}{\Gamma(n - k\alpha)} s^k [f_\alpha(s^{-1/\alpha})]^{-1}$$

$$\int_0^1 p^{n-1-k\alpha} f_\alpha((1 - p)s^{-1/\alpha}) dp \prod_{j=1}^{k} (1 - \alpha)_{n_j-1}. \tag{3}$$

The Poisson–Kingman (ρ_α, γ) class contains some noteworthy subclasses like the *two-parameter* (α, θ) *Poisson–Dirichlet model* [23, 28] which arises for $\gamma_{\alpha,\theta}(t) = \frac{\Gamma(\theta+1)}{\Gamma(\theta/\alpha+1)} t^{-\theta} f_\alpha(t)$ for $0 < \alpha < 1$ and $\theta > -\alpha$, and the *normalized generalized*

Gamma model [26] which arises for $\gamma(t) = \exp\{\psi_\alpha(\lambda) - \lambda t\} f_\alpha(t)$, for $\psi_\alpha(\lambda) = (2\lambda)^\alpha$, $\lambda > 0$, the Laplace exponent of the α Stable density. Notice that those two *PK* models correspond, respectively, to a *polynomial* and an *exponential* tilting of the stable density.

Recently exchangeable Gibbs partitions have been exploited in Bayesian non-parametrics both to identify random probability measure almost surely discrete to be used as alternative priors to the classical Ferguson–Dirichlet model [10] (see [1,18]), like to propose a Bayesian nonparametric approach to species richness estimation in sampling from populations of different species. Here we contribute to those topics by providing some additional finite sample and asymptotic results for *conditional* Gibbs partitions models, which arise when sampling from an almost surely discrete random probability measure $P(\cdot)$ starts after a given initial allocation of n elements in k subgroups.

2 Conditional Gibbs Partitions

Let $(n_1, \ldots, n_k)$ be the vector of multiplicities of k different types observed in an initial sample of n observations from a population of species and assume to be interested in the random allocation of an additional sample of m observations, which may belong to the species already observed or to *new* different species. Under a general Gibbs model (1) the conditional probability to observe a subset of $m - s$ integers *allocated* in the k *old* blocks in configuration $(m_1, \ldots, m_k)$, and the remaining s new observations *partitioned* in $k^* \in (1, m)$ *new* blocks in configuration $(s_1, \ldots, s_{k^*})$ may be derived by the theory of sequential construction of exchangeable Gibbs partitions ([24], see also [3]) and takes the form

$$p_{\alpha, V_{n,k}}(s_1, \ldots, s_{k^*}, m_1, \ldots, m_k | n_1, \ldots, n_k)$$

$$= \frac{V_{n+m, k+k^*}}{V_{n,k}} \prod_{j=1}^{k} (n_j - \alpha)_{m_j} \prod_{j=1}^{k^*} (1 - \alpha)_{s_j - 1}, \quad (4)$$

for $s = \sum_{j=1}^{k^*} s_j \in [0, m]$, $k^* \in [0, s]$ and $\sum_{j=1}^{k} m_j = (m - s)$. Notice that (4) is not invariant to permutations of the arguments, thus conditional Gibbs partitions are not exchangeable. Nevertheless the restriction to the *new* blocks,

$$p_{\alpha, V_{n,k}}(s_1, \ldots, s_{k^*} | n_1, \ldots, n_k) = (n - k\alpha)_{m-s} \frac{V_{n+m, k+k^*}}{V_{n,k}} \prod_{j=1}^{k^*} (1 - \alpha)_{s_j - 1}, \quad (5)$$

does produce an exchangeable partition of the first s integers. In the next Proposition we provide the general form of the joint marginal arising from (4) for $(M_1, \ldots, M_k, S_m)$, for S_m the total number of observations in new blocks, which appears to be new in this setting. We state it without proof since it easily arises by summing over all partitions of $[s]$ elements in k^* blocks and over $k^* \in [0, s]$, and recalling that the number of ways to perform the first operation for the combinatorial structure $(1 - \alpha)_\bullet$ is given by generalized Stirling numbers $S_{s,k*}^{-\lambda_1 -\alpha}$. (See [15,27]).

Proposition 1. *Under the general exchangeable Gibbs partition model (1), given the allocation of n integers in k blocks in configuration $(n_1, \ldots, n_k)$, the probability to observe $m - s$ additional new observations falling in the k old blocks with multiplicities $(m_1, \ldots, m_k)$ and the remaining s observations in new blocks is given by*

$$\mathbb{P}_{\alpha, V_{n,k}}(M_1 = m_1, \ldots, M_k = m_k, S_m = s | n_1, \ldots, n_k)$$

$$= \frac{m!}{s! \prod_{j=1}^{k} m_j!} \prod_{j=1}^{k} (n_j - \alpha)_{m_j} \sum_{k^*=1}^{s} \frac{V_{n+m,k+k^*}}{V_{n,k}} S_{s,k*}^{-1,-\alpha}$$

(6)

for $S_{s,k*}^{-1,-\alpha}$ generalized Stirling numbers defined as the connection coefficients $(x)_s = \sum_{k^*=0}^{s} S_{s,k*}^{-1,-\alpha}(x)_{k^* \uparrow \alpha}$, where $(a)_{b \uparrow c}$ stands for generalized rising factorials $(a)_{b \uparrow c} = a(a + c)(a + 2c) \cdots (a + (b - 1)c)$.

Notice that under the two-parameter (α, θ) Poisson–Dirichlet model, by the specific form of the weights $V_{n,k} = (\theta + \alpha)_{k-1 \uparrow \alpha}/(\theta + 1)_{n-1}$, (6) turns out to be a Multivariate Polya (or compound Dirichlet Multinomial) distribution of parameters $(m, n_1 - \alpha, \ldots, n_k - \alpha, \theta + k\alpha)$, arising by mixing a Multinomial model by a $\mathrm{Dir}(n_1 - \alpha, \ldots, n_k - \alpha, \theta + k\alpha)$ distribution. Further marginalizing (6) with respect to $(M_1, \ldots, M_k)$ yields a Beta-Binomial distribution $(m, \theta + k\alpha, n - k\alpha)$ for the random number S_m of new observations falling in new blocks, as first highlighted in [4].

The following result arises as a consequence of Proposition 1 from the theory in Sect. 3.7 in [24] applied to the $\mathrm{PK}(\rho_\alpha, h \times f_\alpha)$ Gibbs class.

Proposition 2. *Let (X_n) be a sample from a random probability measure $P(\cdot) = \sum_{j=1}^{\infty} P_j \delta_{X_j}(\cdot)$ whose ranked atoms follow a $\mathrm{PK}(\rho_\alpha, h \times f_\alpha)$ distribution, then, given the partition $(n_1, \ldots, n_k)$ induced by the k distinct values in $(X_1, \ldots, X_n)$, the posterior random probability measure admits the following infinite sum representation*

$$P(\cdot)|(n_1, \ldots, n_k) = \sum_{j=1}^{k} \tilde{P}_{j,n} \delta_{\tilde{X}_j}(\cdot) + \tilde{R}_k P_k(\cdot),$$

where (a) $(\tilde{P}_{1,n}, \ldots, \tilde{P}_{k,n}, \tilde{R}_k)$ is the limit in distribution, for $m \to \infty$, of the relative frequencies

$$\left(\frac{M_1}{m}, \ldots, \frac{M_k}{m}, \frac{S_m}{m} \right)$$

for $(M_1, \ldots, M_k, S_m)$ distributed as in (6), (b) P_k is a random probability measure with ranked atoms having distribution $\mathrm{PK}(\rho_\alpha, \gamma_k)$ for $\gamma_k(t)$ the law of $T_k = T\tilde{R}_k$, $T \sim h \times f_\alpha$, and $\tilde{R}_k$ the limit in distribution of $(S_m/m)|K_n = k$ for (cfr. [21])

$$\mathbb{P}_{\alpha, V_{n,k}}(S_m = s | K_n = k) = \binom{m}{s}(n - k\alpha)_{m-s} \sum_{k^*=1}^{s} \frac{V_{n+m,k+k^*}}{V_{n,k}} S_{s,k^*}^{-1,-\alpha}.$$

Notice that for the two-parameter $\mathrm{PK}(\rho_\alpha, \gamma_{\alpha,\theta})$ model Proposition 2 provides the result in Corollary 20 in [24], $P_k(\cdot)$ having distribution $\mathrm{PD}(\alpha, \theta + k\alpha)$ in this case, due to the *deletion of classes property* of the two-parameter model (cfr. [26], Proposition 12; [12]) independently of the vector $(\tilde{P}_{1,\mathbf{n}}, \ldots, \tilde{P}_{k,\mathbf{n}}, \tilde{R}_k)$ having $\mathrm{Dir}(n_1 - \alpha, \ldots, n_k - \alpha, \theta + k\alpha)$ distribution. It is also worth to notice that for normalized generalized Gamma models with $\gamma(t) = [\psi_\alpha(\lambda) - \lambda(t)]f_\alpha(t)$, Proposition 2 can be seen as a reformulation of a result in [16] on the posterior distribution of normalized random measures with independent increments. This is due to a characterization of this model as the unique random probability measure, arising by normalization, inducing random partitions in Gibbs product form (see [2,22]). An explicit result for the *normalized inverse Gaussian model*, corresponding to the generalized Gamma case for $\alpha = 1/2$, has been recently obtained in [9] by deriving the corresponding stick-breaking construction of the *size-biased* (in order of appearance) random atoms.

3 Conditional α Diversity Under $\mathrm{PK}(\rho_\alpha, \gamma)$ Models

The general finite sample distribution of K_n, the number of blocks in an exchangeable Gibbs partition of the first n positive integers, is given by [13]

$$\mathbb{P}(K_n = k) = V_{n,k} S_{n,k}^{-1,-\alpha}.$$

The concept of α-*diversity* for $\mathrm{PK}(\rho_\alpha, h \times f_\alpha)$ partitions was first introduced in [26] (cfr. Sect. 6.1 Proposition 13) as the random variable S_α, with $0 < S_\alpha < \infty$, such that, almost surely, for $n \to \infty$,

$$\frac{K_n}{n^\alpha} \longrightarrow S_\alpha.$$

Pitman shows that $S_\alpha \stackrel{d}{=} T^{-\alpha}$ where $T \sim h \times f_\alpha$. Now, given $K_n = k$ the number of blocks induced by a *basic* sample $(X_1, \ldots, X_n)$, the distribution of K_m, the unknown number of *new* species induced by an additional sample of m observations $(X_{n+1}, \ldots, X_{n+m})$, has been first obtained in [20] and is given by

$$\mathbb{P}_{\alpha, V_{n,k}}(K_m = k^* | K_n = k) = \frac{V_{n+m,k+k^*}}{V_{n,k}} S_{m,k^*}^{-1,-\alpha,-(n-k\alpha)}. \tag{7}$$

Here $S_{m,k^*}^{-1,-\alpha,-(n-k\alpha)}$ are *non-central* generalized Stirling numbers, defined by the convolution relation (see, e.g., (16) in [15])

$$S_{m,k^*}^{-1,-\alpha,-(n-k\alpha)} = \sum_{s=k^*}^{m} \binom{m}{s} (n-k\alpha)_{m-s} S_{s,k^*}^{-1,-\alpha}.$$

Interest in *conditional* α-*diversity* emerges in posterior species richness estimation by the need to obtain asymptotic interval estimation for $K_m | K_n = k$. It is defined as the random variable $S_\alpha^{n,k}$ such that, almost surely, for $m \to \infty$

$$\frac{K_m}{m^\alpha} \Big| (K_n = k) \to S_\alpha^{n,k}.$$

A first result for the conditional α diversity under *two-parameter Poisson–Dirichlet* priors has been derived in [7], by a technique similar to the one used in the original proof for the unconditional case in [27] (Theorem 3.8). In [8] an analogous technique is used to obtain the result under *normalized generalized Gamma* priors. In [4] a simpler alternative derivation for the two-parameter Poisson–Dirichlet family has been obtained by a decomposition approach exploiting the characterization of those models in terms of the *deletion of classes* property previously recalled. In the following Theorem we derive the general distributional result for the entire Poisson–Kingman PK$(\rho_\alpha, h \times f_\alpha)$. In Examples 1 and 3 we show how the previous particular cases arise easily by this general result.

Theorem 1. *Let Π be a* PK(ρ_α, γ) *partition of $\mathbb{N}$ driven by the stable subordinator for some $0 < \alpha < 1$ and some mixing probability distribution γ on $(0, \infty)$. Without loss of generality assume $\gamma(t) = h(t) f_\alpha(t)$. Fix $n \geq 1$ and a partition $(n_1, \dots, n_k)$ of $[n]$ with k positive box-sizes, if Π has conditional α-diversity $S_{\alpha,h}^{n,k}$ then*

$$f_{n,k}^{h,\alpha}(s) = \frac{h(s^{-1/\alpha}) \tilde{g}_{n,k}^\alpha(s)}{\mathbb{E}_{n,k}^\alpha [h(S^{-1/\alpha})]}, \tag{8}$$

for

$$\tilde{g}_{n,k}^\alpha(s) = \frac{\Gamma(n)}{\Gamma(n-k\alpha)\Gamma(k)} s^{k-1/\alpha-1} \int_0^1 p^{n-1-k\alpha} f_\alpha((1-p)s^{-1/\alpha}) dp$$

the density of the product $Y_{\alpha,k} \times [W]^\alpha$ where $Y_{\alpha,k}$ has density

$$g_{\alpha,k\alpha}(y) = \frac{\Gamma(k\alpha+1)}{\Gamma(k+1)} y^k g_\alpha(y)$$

for $g_\alpha(y) = \alpha^{-1} y^{-1-1/\alpha} f_\alpha(y^{-1/\alpha})$, independently of $W \sim \beta(k\alpha, n-k\alpha)$.

Proof. By the unconditional result in [26], the α-diversity for a general $\gamma(t) = h(t)f_\alpha(t)$ mixing density is the r.v. $S_{\alpha,h}$ with density

$$\gamma(s^{-1/\alpha}) = h(s^{-1/\alpha})f_\alpha(s^{-1/\alpha})\alpha^{-1}s^{-1/\alpha-1}, \tag{9}$$

hence, by Bayes' rule,

$$f_{S_{\alpha,\gamma}}(s|n_1,\ldots,n_k) = \frac{p_{\alpha,\gamma}(n_1,\ldots,n_k|s^{-1/\alpha})\gamma(s^{-1/\alpha})}{\int_0^\infty p_{\alpha,\gamma}(n_1,\ldots,n_k|s^{-1/\alpha})\gamma(s^{-1/\alpha})ds}$$

which by (3) in the Introduction, simplifies to

$$f_{S_{\alpha,h}}(s|K_n = k) = \frac{h(s^{-1/\alpha})s^{k-1/\alpha-1}\int_0^1 p^{n-1-k\alpha}f_\alpha((1-p)s^{-1/\alpha})dp}{\int_0^\infty h(s^{-1/\alpha})s^{k-1/\alpha-1}[\int_0^1 p^{n-1-k\alpha}f_\alpha((1-p)s^{-1/\alpha})dp]ds}.$$

Notice that, by definition of mixed $PK(\rho_\alpha, h \times f_\alpha)$ model, the general weights $V_{n,k,h}$ in the EPPF (1) arise as follows

$$V_{n,k,h} = \frac{\alpha^{k-1}}{\Gamma(n-k\alpha)}\int_0^\infty h(s^{-1/\alpha})s^{k-1/\alpha-1}\int_0^1 p^{n-1-k\alpha}f_\alpha((1-p)s^{-1/\alpha})dpds,$$

hence the normalizing constant in formula (8) may be obtained through the following relationship (see also [14])

$$\mathbb{E}_{n,k}^\alpha[h(S^{-1/\alpha})] = V_{n,k,h}\frac{\alpha^{1-k}\Gamma(n)}{\Gamma(k)}. \tag{10}$$

$\square$

Remark 1. The result in Theorem 1 has been obtained independently by an analogous result for the conditional distribution of $T|K_n = k$ first derived in an unfinished manuscript by Ho, James, Lau (Explicit Gibbs Chinese Restaurant Process priors. Personal communication, 2008) that we received by one of those authors as a personal communication. Their result relies on the r.v.

$$R_{\alpha,(n,k)} \overset{d}{=} \frac{S_{\alpha,k\alpha}}{\beta(k\alpha, n-k\alpha)},$$

for $S_{\alpha,k\alpha}$ the polynomially tilted stable random variable with density

$$f_{S_{\alpha,k\alpha}}(t) = \frac{\Gamma(k\alpha+1)}{\Gamma(k+1)}t^{-k\alpha}f_\alpha(t)$$

independent of the $\beta(k\alpha, n-k\alpha)$. It is an easy task to show that $[R_{\alpha,(n,k)}]^{-\alpha} \overset{d}{=} [Y_{\alpha,k} \times \beta(k\alpha, n-k\alpha)^\alpha]$.

Example 1 (Two-parameter Poisson–Dirichlet (α, θ) model). To apply the general result in Theorem 1 recall that the two-parameter Poisson–Dirichlet (α, θ) partition model corresponds to a mixed Poisson–Kingman model with

$$\gamma_{\alpha,\theta}(t) = h(t) \times f_\alpha(t) = \frac{\Gamma(\theta + 1)}{\Gamma(\theta/\alpha + 1)} t^{-\theta} f_\alpha(t),$$

and that the weights in the Gibbs representation of the EPPF can be written as

$$V_{n,k}^{\alpha,\theta} = \alpha^{k-1} \frac{\Gamma(\theta/\alpha + k)\Gamma(\theta + 1)}{\Gamma(\theta + n)\Gamma(\theta/\alpha + 1)}.$$

Hence, by (10), the denominator in (8) corresponds to

$$\mathbb{E}_{n,k}^\alpha(h(Z^{-1/\alpha})) = \mathbb{E}_{n,k}^\alpha(Z^{\theta/\alpha}) = \frac{\Gamma(\theta/\alpha + k)\Gamma(\theta + 1)}{\Gamma(\theta + n)\Gamma(\theta/\alpha + 1)} \frac{\Gamma(n)}{\Gamma(k)},$$

and by Theorem 1 the conditional α diversity $Z_{n,k}^{\alpha,\theta}$ has density

$$f_{n,k}^{\alpha,\theta}(z) = \frac{\Gamma(\theta + n)}{\Gamma(n - k\alpha)\Gamma(\theta/\alpha + k)} z^{\theta/\alpha + k - 1 - 1/\alpha} \int_0^1 f_\alpha[(zw^{-\alpha})^{-1/\alpha}](1 - w)^{n-k\alpha-1} dw \tag{11}$$

which corresponds to the θ/α polynomial tilting of $\tilde{g}_{n,k}^\alpha$. It is an easy task to verify that this is the density of the product of independent r.v.s $Z_{n,k}^{\alpha,\theta} = Y_{\alpha,\theta/\alpha+k} \times [\beta(\theta + k\alpha, n - k\alpha]^\alpha$, as already established in [4]. In the following Proposition, for completeness, we provide an explicit proof that this result agrees with the alternative scale mixture representation $Y_{\alpha,(\theta+n)/\alpha} \times \beta(\theta/\alpha + k, n/\alpha - k)$ obtained in [7] by showing the two decompositions have the same characteristic function.

Proposition 3. *Let $H = Y_1 \times X$ for Y_1 and X independent r.v.s, $Y_1 \sim g_{\alpha,(\theta+n)}$ and $X \sim \beta(\theta/\alpha + k, n/\alpha - k)$, then the r.v. $Z_{n,k}^{\alpha,\theta}$ with density (11) and H have the same characteristic function*

$$G_{\alpha,\theta}^{n,k}(t) = \sum_{r \geq 0} \frac{(it)^r}{r!} \left(\frac{\theta + k\alpha}{\alpha}\right)_r \frac{1}{(\theta + n)_{r\alpha}}.$$

Proof. First notice that by Proposition 2 in [4] (see also [7]), for $m \to \infty$,

$$\mathbb{E}_{\alpha,\theta} \left(\frac{K_m^r}{m^{r\alpha}} \middle| K_n = k\right) \longrightarrow \left(\frac{\theta + k\alpha}{\alpha}\right)_r \frac{\Gamma(\theta + n)}{\Gamma(\theta + n + r\alpha)}.$$

By the change of variable $w = (z/s)^{1/\alpha}$ (11) may be written as

$$f_{n,k}^{\alpha,\theta}(z) = \frac{\Gamma(\theta+n)\Gamma(\theta+k\alpha+1)}{\Gamma(\theta+k\alpha)\Gamma(n-k\alpha)\Gamma((\theta+k\alpha)/\alpha+1)} \frac{1}{\alpha} z^{\theta/\alpha+k-1}$$

$$\times \int_z^\infty \alpha^{-1} s^{-1/\alpha-1} f_\alpha(s^{-1/\alpha}) \left(1-(z/s)^{1/\alpha}\right)^{n-k\alpha-1} ds.$$

Its characteristic function is given by

$$G_{n,k}^{\alpha,\theta}(t) = \frac{\Gamma(\theta+k\alpha+1)}{\Gamma((\theta+k\alpha)/\alpha+1)} \frac{1}{\alpha} \int_z^\infty g_\alpha(s)$$

$$\times \int_0^\infty \exp\{itz\} \frac{\Gamma(\theta+n)}{\Gamma(\theta+k\alpha)\Gamma(n-k\alpha)} z^{\theta/\alpha+k-1} \left(1-(z/s)^{1/\alpha}\right)^{n-k\alpha-1} dzds,$$

and one more change of variable $(z/s)^{1/\alpha} = y$, $z = y^\alpha s$, $dz = s\alpha y^{\alpha-1} dy$ yields

$$= \frac{\Gamma(\theta+k\alpha+1)}{\Gamma((\theta+k\alpha)/\alpha+1)} \frac{1}{\alpha} \int_0^\infty g_\alpha(s)$$

$$\times \int_0^s e^{ity^\alpha s} \frac{\Gamma(\theta+n)}{\Gamma(\theta+k\alpha)\Gamma(n-k\alpha)} (y^\alpha s)^{\theta/\alpha+k-1} (1-y)^{n-k\alpha-1} s\alpha y^{\alpha-1} dyds$$

that reduces to

$$= \frac{\Gamma(\theta+k\alpha+1)}{\Gamma((\theta+k\alpha)/\alpha+1)} \int_0^\infty s^{\theta/\alpha+k} g_\alpha(s)$$

$$\times \int_0^1 e^{ity^\alpha s} \frac{\Gamma(\theta+n)}{\Gamma(\theta+k\alpha)\Gamma(n-k\alpha)} (y)^{\theta+k\alpha-1} (1-y)^{n-k\alpha-1} dyds.$$

Exploiting the known characteristic function of Y^α for $Y \sim \text{Beta}(\theta+k\alpha, n-k\alpha)$ we can write

$$= \frac{\Gamma(\theta+k\alpha+1)}{\Gamma((\theta+k\alpha)/\alpha+1)} \sum_{r=0}^\infty \frac{(it)^r}{r!} \frac{(\theta+k\alpha)_{r\alpha}}{(\theta+n)_{r\alpha}} \int_0^\infty s^{\theta/\alpha+k+r} g_\alpha(s) ds$$

and, by $g_{\alpha,\theta}(z) := [\Gamma(\theta+1)/\Gamma(\theta/\alpha+1)] z^{\theta/\alpha} g_\alpha(z)$,

$$= \sum_{r=0}^\infty \frac{(it)^r}{r!} \frac{(\theta+k\alpha)_{r\alpha}}{(\theta+n)_{r\alpha}} \frac{\Gamma(\theta+k\alpha+1)}{\Gamma((\theta+k\alpha)/\alpha+1)} \frac{\Gamma((\theta+k\alpha+r\alpha)/\alpha+1)}{\Gamma(\theta+k\alpha+r\alpha+1)}. \tag{12}$$

By the usual properties of Gamma function the last expression simplifies to

$$\sum_{r=0}^\infty \frac{(it)^r}{r!} \left(\frac{\theta+k\alpha}{\alpha}\right)_r \frac{1}{(\theta+n)_{r\alpha}}, \tag{13}$$

and the conclusion follows by the result in Proposition 2 in [7] that shows (13) is the characteristic function of $H = Y_1 \times X$.   □

Example 2 (Normalized generalized Gamma model). As previously recalled normalized generalized Gamma partitions models belong to the Poisson–Kingman family for

$$\gamma_{\alpha,\lambda}(t) = \exp\{\psi_\alpha(\lambda) - \lambda t\} f_\alpha(t),$$

where $\psi_\alpha(\lambda) = (2\lambda)^\alpha$, for $\lambda > 0$, is the Laplace exponent of $f_\alpha(\cdot)$. By an application of (9), after the reparametrization $\lambda = \beta^{1/\alpha}/2$, the *unconditional α diversity* for this model (see also [1, 19]) is given by

$$\gamma_{\alpha,\beta}(s^{-1/\alpha}) = \exp\left\{\beta - \frac{1}{2}\left(\frac{\beta}{s}\right)^{1/\alpha}\right\} f_\alpha(s^{-1/\alpha})\alpha^{-1}s^{-1/\alpha-1}.$$

To obtain the density of the conditional posterior α diversity it is enough to apply (8) which provides

$$f_{n,k}^{\alpha,\beta}(s) = \frac{\exp\left\{\beta - \frac{1}{2}\left(\frac{\beta}{s}\right)^{1/\alpha}\right\} \tilde{g}_{n,k}^\alpha(s)}{\mathbb{E}_{n,k}^\alpha\left[\exp\left\{\beta - \frac{1}{2}\left(\frac{\beta}{S}\right)^{1/\alpha}\right\}\right]}.$$

The denominator arises by (10) and the known expression for the $V_{n,k}$ of $PK(\rho_\alpha, \gamma_{\alpha,\beta})$ models as obtained by Corollary (6) in [26],

$$V_{n,k}^{\alpha,\beta} = \frac{e^\beta 2^n \alpha^k}{\Gamma(n)} \int_0^\infty \lambda^{n-1} \frac{e^{-(\beta^{1/\alpha}+2\lambda)^\alpha}}{(\beta^{1/\alpha} + 2\lambda)^{n-k\alpha}} d\lambda,$$

which rewritten in terms of incomplete Gamma functions by the change of variable $x = (\beta^{1/\alpha} + 2\lambda)^\alpha$, $d\lambda = (2\alpha)^{-1}x^{1/\alpha-1}dx$ yields

$$V_{n,k}^{\alpha,\beta} = \frac{e^\beta \alpha^{k-1}}{\Gamma(n)} \sum_{i=0}^{n-1} \binom{n-1}{i}(-1)^i (\beta)^{i/\alpha} \Gamma\left(k - \frac{i}{\alpha}; \beta\right).$$

By (10)

$$\mathbb{E}_{n,k}^\alpha\left[\exp\left\{\beta - \frac{1}{2}\left(\frac{\beta}{S}\right)^{1/\alpha}\right\}\right] = \frac{e^\beta}{\Gamma(k)} \sum_{i=0}^{n-1} \binom{n-1}{i}(-1)^i (\beta)^{i/\alpha} \Gamma\left(k - \frac{i}{\alpha}; \beta\right)$$

hence by Theorem 1 the conditional α diversity $S_{n,k}^{\alpha,\beta}$ for the normalized generalized Gamma model has density

$$f_{n,k}^{\alpha,\beta}(s) = \frac{\Gamma(k)\exp(2^{-1}(\beta/s)^{1/\alpha})\tilde{g}_{n,k}^{\alpha}(s)}{\sum_{i=0}^{n-1}\binom{n-1}{i}(-1)^i(\beta)^{i/\alpha}\Gamma(k-\frac{i}{\alpha};\beta)}.$$

The result agrees with [8] (Theorem 1) due to the equivalence in distribution between $Y_{\alpha,n/\alpha} \times \beta(k, n/\alpha - k)$ and $Y_{\alpha,k} \times [\beta(k\alpha, n - k\alpha)]^{\alpha}$ which follows as a particular case ($\theta = 0$) of Proposition 3.

4 Stirling's Approximation of Exchangeable Gibbs Weights

The computation of the $V_{n,k}$ Gibbs weights in the PK(ρ_α, γ) class, apart from the two-parameter (α, θ) case, can be particularly demanding for large values of n and k. Notice that this class contains potentially infinitely many models depending on the particular choice of the mixing density. See [14] for some additional explicit classes beyond the generalized Gamma and the two-parameter cases. Hence it is tempting to investigate if there is any possibility to obtain some sort of approximation for those weights. Here we obtain a preliminary result by combining approximation results for generalized Stirling numbers with the asymptotic results for the law (conditional and unconditional) of the number of blocks treated in the previous section.

Proposition 4. *The following Stirling's approximation holds for large n for the weights $V_{n,k}$ of an exchangeable Gibbs partition with PK($\rho_\alpha, h \times f_\alpha$) distribution*

$$V_{n,k} \approx \frac{\alpha^{k-1}\Gamma(k)}{\Gamma(n)} h\left[\left(\frac{k}{n^\alpha}\right)^{-1/\alpha}\right]. \tag{14}$$

Proof. By a result in [25] (eq. (96)) the following asymptotic formula for the generalized Stirling numbers $S_{n,k}^{-1,-\alpha}$ for $n \to \infty$ and $0 < s < \infty$, with $k \approx sn^\alpha$ is derived by known local limit approximations by the stable density for the number of blocks in a partition generated by a PD(α, α) model

$$S_{n,k}^{-1,-\alpha} \approx \frac{\alpha^{1-k}\Gamma(n)}{\Gamma(k)} g_\alpha(s)n^{-\alpha}, \tag{15}$$

where, as before, $g_\alpha(s) = \alpha^{-1} f_\alpha(s^{-1/\alpha})s^{-1-1/\alpha}$. Now, by the known local limit approximation for the distribution of the number of blocks which follows by the asymptotic distribution of K_n/n^α previously recalled

$$\mathbb{P}(K_n = k) \approx h(s^{-1/\alpha})g_\alpha(s)n^{-\alpha}$$

for $k \approx sn^\alpha$, hence the result follows by substitution. $\qquad\square$

Example 3. It is an easy task to prove that (14) agrees with the approximation which can be obtained directly by the simple expression of the $PD(\alpha, \theta)$ weights,

$$V_{n,k}^{\alpha,\theta} = \frac{\alpha^{k-1}(\theta/\alpha + 1)_{k-1}}{(\theta + 1)_{n-1}} = \frac{\alpha^{k-1}\Gamma(\theta/\alpha + k)\Gamma(\theta + 1)}{\Gamma(\theta + n)\Gamma(\theta/\alpha + 1)},$$

multiplying and dividing by $\Gamma(k)\Gamma(n)$ and by first order Stirling's approximation $\Gamma(n + a)/\Gamma(n + b) = n^{a-b}$ yields

$$V_{n,k}^{\alpha,\theta} \approx \frac{\alpha^{k-1}\Gamma(k)}{\Gamma(n)}\left(\frac{k}{n^\alpha}\right)^{\theta/\alpha}\frac{\Gamma(\theta + 1)}{\Gamma(\theta/\alpha + 1)}.$$

By the same technique applied to *non-central* generalized Stirling numbers an analogous Stirling's approximation result holds for *conditional* Gibbs weights:

Proposition 5 ([5]). *The following Stirling's approximation holds for large m for the conditional weights* $V_{n+m,k+k^*}/V_{n,k}$ *of an exchangeable Gibbs partition with* $PK(\rho_\alpha, h \times f_\alpha)$ *distribution*

$$\frac{V_{n+m,k+k^*}}{V_{n,k}} \approx \frac{h(s^{-1/\alpha})s^k\alpha^{k^*}\Gamma(k^*)\Gamma(n)m^{-(n-k\alpha)}}{\mathbb{E}_{n,k,\alpha}(h(S^{-1/\alpha}))\Gamma(m)\Gamma(k)}$$

$$= \frac{\alpha^{k+k^*-1}h(s^{-1/\alpha})s^k\Gamma(k^*)m^{-(n-k\alpha)}}{V_{n,k}\Gamma(m)}. \tag{16}$$

Acknowledgments The author is partially supported by grant PRIN MIUR:2008CEFF37.

References

1. Cerquetti, A.: A note on Bayesian nonparametric priors derived from exponentially tilted Poisson-Kingman models. Stat. Probab. Lett. **77**(18), 1705–1711 (2007)
2. Cerquetti, A.: On a Gibbs characterization of normalized generalized Gamma processes. Stat. Probab. Lett. **78**, 3123–3128 (2008)
3. Cerquetti, A.: A generalized sequential construction of exchangeable Gibbs partitions with application. In: Proceedings of S.Co. 2009. Milano, Italy (2009)
4. Cerquetti, A.: A decomposition approach to Bayesian nonparametric estimation under two-parameter Poisson-Dirichlet priors. In: Proceedings of ASMDA 2011, Rome, Italy (2011)
5. Cerquetti, A.: Stirling's approximations for exchangeable Gibbs weights. arXiv:1206.6812v1 [math.PR] (2012)
6. Ewens, W.J.: The sampling theory of selectively neutral alleles. Theor. Popul. Biol. **3**, 87–112 (1972)
7. Favaro, S., Lijoi, A., Mena, R.H., Prünster, I.: Bayesian non-parametric inference for species variety with a two-parameter Poisson-Dirichlet process prior. J. Roy. Stat. Soc. B **71**, 993–1008 (2009)
8. Favaro, S., Lijoi, A., Prünster, I.: Asymptotics for a Bayesian nonparametric estimator of species variety. **18**, 1267–1283 Bernoulli (2012a)

9. Favaro, S., Lijoi, A., Prünster, I.: On the stick breaking representation of normalized inverse Gaussian priors. **9**, 663–674 Biometrika (2012b)
10. Ferguson, T.S.: A Bayesian analysis of some nonparametric problems. Ann. Stat. **1**, 209–230 (1973)
11. Fisher, R.A., Corbet, A.S., Williams, C.B.: The relation between the number of species and the number of individuals in a random sample of an animal population. J. Animal Ecol. **12**, 42–58 (1943)
12. Gnedin, A., Haulk, S., Pitman, J.: Characterizations of exchangeable partitions and random discrete distributions by deletion properties. arXiv:0909.3642 [math:PR] (2009)
13. Gnedin, A., Pitman, J.: Exchangeable Gibbs partitions and Stirling triangles. J. Math. Sci. **138**(3), 5674–5685 (2006)
14. Ho, M.-W., James, L.F., Lau, J.W.: Gibbs partitions (EPPF's) derived from a stable subordinator are Fox H – And Meijer G – Transforms. arXiv:0708.0619v2 [math.PR] (2007)
15. Hsu, L.C., Shiue, P.J.: A unified approach to generalized Stirling numbers. Adv. Appl. Math. **20**, 366–384 (1998)
16. James, L.F., Lijoi, A., Prünster, I.: Posterior analysis for normalized random measures with independent increments. Scand. J. Stat. **36**, 76–97 (2009)
17. Kingman, J.F.C.: The representation of partition structures. J. Lond. Math. Soc. **18**, 374–380 (1978)
18. Lijoi, A., Mena, R.H., Prünster, I.: Hierarchical mixture modeling with normalized Inverse Gaussian priors. J. Am. Stat. Assoc. **100**, 1278–1291 (2005)
19. Lijoi, A., Mena, R.H., Prünster, I.: Controlling the reinforcement in Bayesian non-parametric mixture models. J. Roy. Stat. Soc. B **69**(4), 715–740 (2007a)
20. Lijoi, A., Mena, R.H., Prünster, I.: Bayesian nonparametric estimation of the probability of discovering new species. Biometrika **94**, 769–786 (2007b)
21. Lijoi, A., Prünster, I., Walker, S.G.: Bayesian nonparametric estimators derived from conditional Gibbs structures. Ann. Appl. Probab. **18**, 1519–1547 (2008a)
22. Lijoi, A., Prünster, I., Walker, S.G.: Investigating nonparametric priors with Gibbs structure. Statistica Sinica **18**, 1653–1668 (2008b)
23. Pitman, J.: Exchangeable and partially exchangeable random partitions. Probab. Theor. Rel. Fields **102**, 145–158 (1995)
24. Pitman, J.: Some developments of the Blackwell-MacQueen urn scheme. In: Ferguson, T.S., Shapley, L.S., MacQueen, J.B. (eds.) Statistics, Probability and Game Theory. IMS Lecture Notes-Monograph Series, vol. 30, pp. 245–267. Institute of Mathematical Statistics, Hayward (1996)
25. Pitman, J.: Brownian motion, bridge, excursion, and meander characterized by sampling at uniform times. Elect. J. Prob. **4**, 1–33 (1999)
26. Pitman, J.: Poisson-Kingman partitions. In: Goldstein, D.R. (ed.) Science and Statistics: A Festschrift for Terry Speed. Lecture Notes-Monograph Series, vol. 40, pp. 1–34. Institute of Mathematical Statistics, Hayward (2003)
27. Pitman, J.: Combinatorial stochastic processes. Ecole d'Eté de Probabilité de Saint-Flour XXXII – 2002. Lecture Notes in Mathematics N. 1875. Springer, Berlin (2006)
28. Pitman, J., Yor, M.: The two-parameter Poisson-Dirichlet distribution derived from a stable subordinator. Ann. Probab. **25**, 855–900 (1997)

Estimation of Traffic Matrices for LRD Traffic

Pier Luigi Conti, Livia De Giovanni, and Maurizio Naldi

Abstract The estimation of traffic matrices in a communications network on the basis of a set of traffic measurements on the network links is a well-known problem, for which a number of solutions have been proposed when the traffic does not show dependence over time, as in the case of the Poisson process. However, extensive measurements campaigns conducted on IP networks have shown that the traffic exhibits long-range dependence (LRD). Here two methods are proposed for the estimation of traffic matrices in the case of LRD, their asymptotic properties are studied, and their relative merits are compared.

1 Introduction

Traffic matrices play a fundamental role in network management, because they are used to describe the amount of bits (packets) transmitted from every Source to every Destination pair (S–D pair, for short). Their importance comes from a basic property [1]: they are invariant under changes of either the network topology or routing.

Formally speaking, a telecommunication network can be represented as a graph, where nodes are transmission devices and arcs are (physical) links connecting nodes. In a network with n nodes there are typically (at most) $N = n(n - 1)$ S–D

P.L. Conti (✉)
Department of Statistical Sciences, Sapienza University of Rome, Rome, Italy
e-mail: pierluigi.conti@uniroma1.it

L. De Giovanni
Department of Political Science, LUISS University, Rome, Italy
e-mail: ldegiovanni@luiss.it

M. Naldi
Department of Computer Science and Civil Engineering, University "Tor Vergata" of Rome, Rome, Italy
e-mail: naldi@disp.uniroma2.it

M. Grigoletto et al. (eds.), *Complex Models and Computational Methods in Statistics*, Contributions to Statistics, DOI 10.1007/978-88-470-2871-5_8,
© Springer-Verlag Italia 2013

pairs, but only M links, with M considerably smaller than N. As a consequence, traffic on links does not identify S–D traffic. In other words, information produced by observations on links does not allow one to identify the S–D traffic. This means we are facing with an incomplete information (or, equivalently, an underconstrained problem).

Approaches to the estimation of S–D traffic matrices when only traffic flowing on links is observed, are essentially two. On the one hand, engineering literature has first considered techniques based on numerical optimization; good reviews are in the papers [11, 15]. On the other hand, since the seminal paper by [30], the statistical approach has received an increasing attention. Statistical approach is based either on the maximum likelihood method [2, 3, 30], or on Bayesian methods [29, 32]. In [3] a functional mean variance relation of S–D traffic ensuring the identifiability of the model (under special assumptions on the network topology) is introduced. A comprehensive account of the statistical literature is in [4].

In the above-mentioned papers, S–D pairs are assumed to behave independently. Furthermore, S–D traffic counts for each S–D pair are assumed either Poisson or Gaussian, as well as independent over successive measurements periods. The basic assumptions considered in the literature are listed below.

(a) S–D pairs are independent.
(b) The traffic produced by a single S–D pair is stationary (either Gaussian or Poisson).
(c) The traffic produced in different time intervals by an S–D pair is independent.

Assumptions (a), (b) are validated in different papers. In detail, independence of S–D pairs is empirically confirmed in [27], where data coming from the Finnish university network (Funet) are analyzed. In particular, empirical correlations between standardized residuals of bits (packets) arrival process (bit/packet network traffic) for different S–D pairs, at various time aggregation levels, show that the independence assumption for S–D pairs can be reasonably assumed, even for S–D pairs sharing the same source node or the same destination node.

As far as the traffic counts probability distribution is concerned, the assumption of Poisson traffic is hardly ever used, and even in the seminal paper by [30], a Gaussian approximation is considered. Gaussianity assumption essentially rests on the central limit theorem: S–D traffic is produced by the superposition of several independent elementary sources alternating ON and OFF periods (ON–OFF sources, for short), so that it tends to be Gaussian as the number of ON–OFF sources increases [24, 28]. The assumption has been considered in [3, 16, 17, 23, 27], and validated via empirical analysis, again on Funet traffic data.

The assumption of stationarity is studied in [3, 23]. The main conclusion is that stationarity can be reasonably assumed to hold within a period of 30–90 min. The empirical studies performed in [16, 17, 27] for Funet data essentially confirm the stationarity assumption, with a time aggregation from 1 s to 300 s.

The most criticizable assumption is (c), namely the independence of traffic generated by a single S–D source in different time intervals. Since the paper by [18], different empirical analyses have shown that the arrival process of (bits) packets is self-similar, with increments characterized by a strong time correlation. More formally, traffic obtained by aggregating independent ON–OFF sources exhibits Long Range Dependence (LRD). A deep theoretical analysis is in [28], where a functional central limit theorem for aggregated traffic is proved. The main results in that paper can be interpreted as follows.

1. The superposition of independent ON/OFF sources with heavy-tailed ON and/or OFF periods produces a limiting Gaussian self-similar aggregate cumulative packets arrival process.
2. The expected traffic level provides the main term of the observed traffic; fluctuations around expected traffic can be approximated by a (rescaled) fractional Brownian motion.
3. If a finite number of independent heterogeneous sources, possibly with different values of the Hurst parameter H, are superimposed, then the term with the highest value of H tends to be dominant. Hence, the application with the highest value of the Hurst parameter determines the value of the Hurst parameter for the whole aggregate traffic. This point is also raised in [10].

Empirical studies to validate the assumption that ON and/or OFF periods in Local Area Networks have a heavy-tailed distribution are in [25, 34, 35]. Similar studies have been performed for Wide Area Networks to validate the assumption that session durations possess a heavy-tailed probability distribution; see [9, 25].

The presence of LRD of aggregate traffic is detected in [16, 27] for Funet data. See also [26], again with a time aggregation from 10 ms to 60 s, where a wavelet analysis (as developed in [33]) is performed. In [23] LRD is detected through a visual analysis. All the above-mentioned papers show that S–D traffic is characterized by the presence of LRD (with values of the Hurst parameter H ranging between 0.65 and 0.9).

Motivated by theoretical results (1)–(3), as well as by the above-mentioned empirical analyses, Gaussian traffic models explicitly accounting for the presence of LRD have been considered in [5–7]. In [5, 6] maximum likelihood estimation based on EM algorithm is studied. In [7] the likelihood function for link traffic data is considered, and asymptotics for the roots of the likelihood equations are studied. Comparisons with the main results in [5,6] are performed, and an application to real data (again showing the presence of LRD) is provided.

In the sequel the model introduced in [7] will be shortly presented, and the main results will be reviewed. Motivated by computational reasons, a form of pseudo-likelihood will be further introduced, and the properties of the corresponding maximum pseudo-likelihood estimators will be studied. Comparisons with maximum likelihood estimators will be further considered.

2 A Gaussian Model for LRD

Let us denote by X_i^t the amount of traffic for the S–D pair i at time-slot t, and by $X^t = (X_1^t, \ldots, X_N^t)$ the vector of traffic for all N S–D pairs at time t. The assumptions on which our model rests are listed below.

1. The stochastic process $(X^t; t \geq 1)$ is a stationary Gaussian process, with:

$$E[X_i^t] = \mu_{Xi}, \quad V[X_i^t] = \sigma_{Xi}^2, \quad i = 1, \ldots, N; \quad t \geq 1.$$

2. Different S–D pairs generate independent traffic:

$$C[X_i^t, X_j^{t+k}] = 0, \quad i \neq j, \quad i, j = 1, \ldots, N; \quad t \geq 1, \quad k \geq 0.$$

3. The auto-correlation function of lag τ for S–D pairs possesses the form:

$$C[X_i^t, X_i^{t+\tau}] = \sigma_{Xi}^2 \frac{1}{2} \{ (\tau + 1)^{2H} - 2\tau^{2H} + (\tau - 1)^{2H} \}$$

$$= \sigma_{Xi}^2 \, \rho_X(\tau), \quad i = 1, \ldots, N; \quad \tau \geq 0$$

where $1/2 \leq H < 1$ is the *Hurst* parameter. When $H = 1/2$, the correlation between X_i^t and $X_i^{t+\tau}$ is zero; this is the *Short Range Dependence* case. When $H > 1/2$, the correlation between X_i^t and $X_i^{t+\tau}$ is non-null, and slowly decreasing as τ increases. This is the LRD case.

The spectral function for a single S–D pair is equal to:

$$f_{X_i}(\omega; H) = \sigma_{Xi}^2 \, f_X(\omega; H)$$

$$= \sigma_{Xi}^2 \, \frac{1}{\pi} \, \sin(\pi H) \, \Gamma(2H + 1) \, (1 - \cos\omega) C_0(H, \omega), \quad i = 1, \ldots, N$$

$$(1)$$

where

$$C_l(H, \omega) = \sum_{k=-\infty}^{+\infty} (\log |2\pi k + \omega|)^l |2\pi k + \omega|^{-(2H+1)}, \quad l = 0, \pm 1, \pm 2, \ldots$$

$$(2)$$

When ω is close to zero, the well-known approximation

$$f_{X_i}(\omega; H) \approx \sigma_{Xi}^2 \, \frac{1}{2\pi} \, \sin(\pi H) \, \Gamma(2H + 1) \, |\omega|^{1-2H} \quad \text{as } \omega \to 0, \quad i = 1, \ldots, N$$

$$(3)$$

holds.

For the sake of simplicity, from now on we will use the following notation:

$$X^t = (X_1^t, \ldots, X_N^t)$$

$$\mu_X = (\mu_{X1}, \ldots, \mu_{XN})$$

$$\theta_i = \sigma_{Xi}^2, \quad i = 1, \ldots, N$$

$$\theta = (\sigma_{X1}^2, \ldots, \sigma_{XN}^2).$$

As a consequence of the independence among S–D pairs, the lag τ covariance matrix of the r.v. X^t is equal to

$$\Gamma_X(\tau; \ \theta, \ H) = \rho_X(\tau) \begin{bmatrix} \theta_1 & 0 & \cdots & 0 \\ 0 & \theta_2 & \cdots & 0 \\ \multicolumn{4}{c}{\dotfill} \\ 0 & 0 & \cdots & \theta_N \end{bmatrix} = \rho_X(\tau) \, \Sigma_X(\theta).$$

Assumptions 1–3 imply that traffic over links is characterized by LRD, too. To see this, let Y_k^t be the traffic flowing on the link k at time slot t, and let $Y^t = (Y_1^t, \ldots, Y_M^t)$ be the (column) vector containing the traffic for the M links. Denote further by $A = (a_{kl})$ the $M \times N$ matrix where a_{kl} equal either to 1 or to 0 according to whether link k does or does not belong to the directed path of the S–D pair l. Y^t and X^t are related by the relationship

$$Y^t = AX^t, \quad t = 1, 2, \ldots \tag{4}$$

The matrix A is referred to in engineering literature as *routing matrix*.

Taking into account (4) and assumptions 1–3, it is seen that the (multivariate) process $(Y^t; \ t \geq 1)$ is a stationary Gaussian process, with mean function $\mu_Y = A\mu_X$ and covariance matrix of lag τ equal to

$$\Gamma_Y(\tau; \ \theta, \ H) = \rho_X(\tau) G(\theta) \tag{5}$$

where $G(\theta)$ is the $M \times M$ matrix

$$G(\theta) = A \Sigma_X(\theta) A'. \tag{6}$$

The cross-spectrum matrix of the process $(Y^t; \ t \geq 1)$ is equal to

$$\Phi_Y(\omega; \ \theta, \ H) = f_X(\omega; \ H) \, G(\theta) \tag{7}$$

where $f_X(\omega; \ H)$ and $G(\theta)$ are given by (1) and (6), respectively. Note that the cross-spectrum matrix (7) factorizes into the product of two terms: a scalar only depending on H and a $M \times M$ matrix only depending on θ.

3 Identifiability Issues

As already said, the main source of trouble is that the sample observations are *not* $X^1, \ldots, X^T$, but $Y^1, \ldots, Y^T$. Since M (the number of links) is usually considerably smaller than N (the number of S–D pairs), statistical data do not generally identify the model. Identifiability issues are dealt with in [7], where results by [3] are reworked and extended. We consider here a short summary of the main results, with a few remarks.

First of all, since M is usually smaller than N, the model introduced so far is not identifiable whenever the parameters μ_X, θ are free. The simplest idea consists in using a mean–variance relationship, such as $\mu_{Xi} = \theta_i$, o, more generally, $\mu_{Xi} = \text{const} \times \theta_i^q$, $q > 0$. In the sequel, we will assume that

$$\mu_{Xi} = h(\theta_i), \quad i = 1, \ldots, N \tag{8}$$

where $h(\cdot)$ is a strictly monotone, known function. From now on, the vector notation

$$h(\theta) = [h(\theta_1) \cdots h(\theta_N)] \tag{9}$$

will be used.

The mean–variance relationships (8) are not enough to make the model identifiable. To this purpose, we need to introduce a further restriction on the topology of the communication network considered.

As already said, a telecommunication network can be seen as a graph, where transmission devices are nodes, and direct links are arcs. A sequence of consecutive arcs connecting two nodes is a *path*. The *length* of a path is the number of arcs defining the path.

A sub-path of a given path is a sub-sequence of the arcs of the path. A sub-path is still a path, connecting two nodes. With a slightly different notation, let us indicate nodes by letters, and arcs by pairs of letters, those of the two nodes connected by arcs. Suppose that two nodes a, b are connected through the path composed by the l arcs $(a, a_1), (a_1, a_2), \ldots, (a_{l-1}, b)$. Then such a path also contains all paths from a node a_i to a node a_j, $0 \leq i < j \leq l$, with $a_0 = a$, $a_l = b$. They are *sub-paths* of the path connecting the nodes a and b. Clearly, every sub-path connects two nodes. In the sequel, we will assume that the paths in the routing matrix possess the following property.

(G)—For every pair $a_i, a_j, i < j$, of nodes, the sub-path $(a_i, a_{i+1}), \ldots, (a_{j-1}, a_j)$ is also the path connecting the Source-node a_i to the Destination-node a_j, as it appears in the routing matrix A.

The simplest algorithm to construct S–D and satisfying (G) is the *minimum length* rule: paths are composed by the smallest number of arcs connecting the Source-node to the Destination-node.

Proposition 1. *Suppose that: (i) the model satisfy conditions 1–3 of Sect. 2; (ii) the mean–variance relationships (8) holds; (iii) all S–D pairs can be connected through a path, for which (G) holds. Then, the model is identifiable if observed data are $Y^1, \ldots, Y^T$.*

Proof. Easy consequence of results in [7].   □

4 Approach Based on the (Approximated) Full Likelihood

4.1 Construction of the Approximate Log-Likelihood

The earlier approach to the estimation of μ_x, θ (see [5, 6]) consists in considering the (unobserved) S–D traffic vectors X^1, ..., X^T as missing data, and to use the EM algorithm based on the observed link traffic vectors Y^1, ..., Y^T.

A more recent and fruitful approach [7] consists in using the full likelihood function based on the observed data Y^1, ..., Y^T. Since the exact construction of the full likelihood is awkward, the main idea is to resort to a multivariate version of the Whittle approximation [13, 36].

The approximate log-likelihood for the unknown parameters (θ, H) is given by

$$\bar{l}_T(\theta, H) = -\sum_{j=1}^{T} \{\log \det(f_X(\omega_j; H) G(\theta))$$

$$+ \operatorname{tr}((f_X(\omega_j; H) G(\theta))^{-1}) \widehat{I}_T(y, \omega_j))\} \tag{10}$$

where $\omega_j = 2\pi j/T - \pi$, and $\widehat{I}_T(\omega)$ is the empirical periodogram, defined as

$$\widehat{I}_T(\omega) = \widehat{w}_T(\omega) \widehat{w}_T(\omega)^*$$

with

$$\widehat{w}_T(\omega) = \frac{1}{\sqrt{2\pi T}} \sum_{t=1}^{T} Y^t e^{it\omega}.$$

In order to (locally) maximize (10) w.r.t. H and θ, we need to compute the corresponding partial derivatives. Using the rules in [20], it is immediately seen that

$$\frac{\partial}{\partial H} \left\{ \sum_{j=1}^{T} \log \det(f_X(\omega_j; H) G(\theta)) \right\}$$

$$= M \sum_{j=1}^{T} \frac{1}{f_X(\omega_j; H)} \frac{1 - \cos \omega_j}{\pi} \{(\pi \cos(\pi H)\Gamma(2H + 1) + 2 \sin(\pi H) \Gamma'(2H + 1))$$

$$\times C_0(H, \omega_j) - 2\sin(\pi H) \Gamma(2H + 1)C_1(H, \omega_j)\} \tag{11}$$

where $\Gamma'(x) = d\Gamma(x)/dx$ and $C_l(H, \omega)$ is given by (2). Similarly, the derivative w.r.t. θ of the log det term appearing in (10) is equal to

$$\frac{\partial}{\partial \theta} \left\{ \sum_{j=1}^{T} \log \det(f_X(\omega_j; H) G(\theta)) \right\} = T \left(\text{vec}((G(\theta)^{-1})')' \frac{\partial G(\theta)}{\partial \theta} \right) \qquad (12)$$

where the matrix $G(\theta)$ is defined in (6), and $\frac{\partial G(\theta)}{\partial \theta}$ is a $M^2 \times N$ matrix having structure:

$$\left[\left(\frac{d \, \text{vec}(G(\theta))}{d \, \sigma_{X1}^2} \right) \cdots \left(\frac{d \, \text{vec}G(\theta)}{d \, \sigma_{XN}^2} \right) \right].$$

As far as the computation of the derivatives of the "tr" term in (10) is concerned, we have

$$\frac{\partial}{\partial H} \left\{ \sum_{j=1}^{T} \text{tr}((f_X(\omega_j; H) G(\theta))^{-1}) \widehat{I}_T(\omega_j)) \right\}$$

$$= - \sum_{j=1}^{T} \text{tr}(G(\theta)^{-1} \widehat{I}_T(\omega_j)) \frac{1}{f_X(\omega_j; H)^2} \frac{1 - \cos \omega_j}{\pi} \{(\pi \cos(\pi H) \Gamma(2H + 1)$$

$$+2 \sin(\pi H) \Gamma'(2H + 1))C_0(H, \omega_j) - 2 \sin(\pi H) \Gamma(2H + 1)C_1(H, \omega_j)\}$$

$$\tag{13}$$

and

$$\frac{\partial}{\partial \theta} \left\{ \sum_{j=1}^{T} \text{tr}((f_X(\omega_j; H) G(\theta))^{-1}) \widehat{I}_T(\omega_j)) \right\}$$

$$= \sum_{j=1}^{T} f_X(\omega_j; H)^{-1} (\text{vec}(\widehat{I}_T(\omega_j)))' \left(\frac{\partial G(\theta)^{-1}}{\partial \theta} \right) \qquad (14)$$

where, as shown in [20] (p. 208, Table 7),

$$\frac{\partial G(\theta)^{-1}}{\partial \theta} = -((G(\theta)')^{-1} \otimes G(\theta)^{-1}) \frac{\partial G(\theta)}{\partial \theta}.$$

If $\widehat{H}, \widehat{\theta}$ denote the (local) maximizers of $\bar{l}_T(\theta, H)$ (10), the mean vector μ_X is then estimated by inverting the relationship $\mu_X = h(\theta)$, i.e. by taking

$$\widehat{\mu}_{Xi} = h^{-1}(\widehat{\theta}_i), \quad i = 1, \ldots, N.$$

4.2   Properties of the MLE Estimators

The asymptotic properties of the estimators $\widehat{H}, \widehat{\theta}$ are obtained in [7]. Here we give a short summary. Let H_0, θ_0 be the "true values" of H, θ, respectively, and let $Q_1(H, \theta), Q_2(H, \theta), Q(H, \theta)$ be the functions:

$$Q_1(H, \theta) = M \int_{-\pi}^{\pi} \log(f_X(\omega, H)) \, d\omega + 2\pi \, \log \det(G(\theta)), \qquad (15)$$

$$Q_2(H, \theta) = \mathrm{tr}(G(\theta)^{-1} G(\theta_0)) \int_{-\pi}^{\pi} \frac{f_X(\omega, H_0)}{f_X(\omega, H)} \, d\omega, \qquad (16)$$

$$Q(H, \theta) = Q_1(H, \theta) + Q_2(H, \theta). \qquad (17)$$

Denote further by W the $(N+1) \times (N+1)$ matrix having elements

$$w_{kl}(H, \theta) = \frac{\partial^2 Q(H, \theta)}{\partial \theta_k \, \partial \theta_l}, \quad k, l = 1, \ldots, N \qquad (18)$$

$$w_{N+1,l}(H, \theta) = w_{l,N+1}(H, \theta) = \frac{\partial^2 Q(H, \theta)}{\partial H \, \partial \theta_l}, \quad l = 1, \ldots, N \qquad (19)$$

$$w_{N+1,N+1}(H, \theta) = \frac{\partial^2 Q(H, \theta)}{\partial H^2}. \qquad (20)$$

If $x = (x_1, \ldots, x_{m^2})$ is a vector of m^2 elements, let $\mathrm{Ma}(x)$ be the $m \times m$ matrix whose hth row is composed by $x_{m(j-1)+1}, \ldots, x_{mj}, j = 1, \ldots, m$. Finally, let $\gamma_k(\omega; H, \theta)$ be the $M \times M$ matrices

$$\gamma_k(\omega; H, \theta) = f_X(\omega, H)^{-1} \left\{ \mathrm{Ma}\left(\frac{\partial G(\theta)^{-1}}{\partial \theta_k} \right) \right\}$$

$$= -f_X(\omega, H)^{-1} \mathrm{Ma}\left((G(\theta)^{-1} \otimes G(\theta)^{-1}) a_k \right), \quad k = 1, \ldots, N \qquad (21)$$

$$\gamma_{N+1}(\omega; H, \theta) = \left(\frac{d f_X(\omega, H)^{-1}}{d H} \right) G(\theta)^{-1} = -\frac{f_X'(\omega, H)}{f_X(\omega, H)^2} G(\theta)^{-1} \qquad (22)$$

and let $V(H, \theta)$ be the $(N+1) \times (N+1)$ matrix with elements

$$v_{kl}(H, \theta) = 4\pi \int_{-\pi}^{\pi} f_X(\omega, H_0)^2 \mathrm{tr} \left(\gamma_k(\omega; H, \theta) G(\theta_0) \gamma_l(\omega; H, \theta) G(\theta_0) \right) d\omega.$$

$$(23)$$

Proposition 2. *If the model satisfies assumptions 1–3, as T increases*

$$\sqrt{T}\begin{bmatrix}\widehat{\theta}-\theta_0\\\widehat{H}-H_0\end{bmatrix}\xrightarrow{d}\mathcal{N}_{N+1}(0,W(H_0,\theta_0)V(H_0,\theta_0)W(H_0,\theta_0)') \qquad (24)$$

$\mathcal{N}_p(\mu,D)$ *denoting a p-variate normal distribution with mean vector μ and covariance matrix D.*

Proof. See [7]. □

The matrices $V(H_0,\theta_0)$, $W(H_0,\theta_0)$ can be consistently estimated. It is enough to define $\widehat{Q}_1(H,\theta)$, $\widehat{Q}_2(H,\theta)$, $\widehat{Q}(H,\theta)$ be defined exactly as (15), (16), (17), with integrals are replaced by Riemann sums evaluated at Fourier frequencies $\omega_j=2\pi j/T-\pi$, $j=1,\ldots,T$, and H_0, θ_0 are replaced by $\widehat{H}$, $\widehat{\theta}$, respectively. Similarly, let $w_{kl}(H,\theta)$, $\widehat{v}_{kl}(H,\theta)$ be defined as in (20), (23), again with integrals replaced by Riemann sums at Fourier frequencies, and let $\widehat{W}(H,\theta)$, $\widehat{V}(H,\theta)$ be the corresponding matrices. In is easy to show that

$$\widehat{W}(\widehat{\theta},\widehat{H})\xrightarrow{p}W(H_0,\theta_0),\ \widehat{V}(\widehat{\theta},\widehat{H})\xrightarrow{p}V(H_0,\theta_0)\ \text{as}\ T\to\infty \qquad (25)$$

so that the asymptotic covariance matrix appearing in (2) can be consistently estimated.

5 Approach Based on Pseudo-Likelihood

5.1 Construction of the Pseudo-Likelihood

The approach based on the full likelihood (although considerably simplified) is computationally heavy, at least for large networks. As it appears from the previous section, differentiating the log-likelihood w.r.t. θ requires the inversion of the $M\times M$ matrix $G(\theta)$ (and the computation of its determinant, as well). Among the most well-known inversion algorithms, Gaussian elimination does have a computational complexity $O(M^3)$, and Strassen's algorithm possesses a $O(M^{\log_2 7})\approx O(M^{2.807})$ complexity. As M gets large (which usually happens for large networks), a super-quadratic computational complexity could be unaffordable.

A computational simplification can be obtained by using a pseudo-likelihood based on pairs of links, instead of the full likelihood. Under the assumption of independence of the traffic produced in different time intervals by an S–D pair, that form of pseudo-likelihood was first proposed in [19]. We adopt here the same idea, but when traffic is characterized by the presence of LRD. Let $s=(k,h)$ be a pair of link, and define the vector (of dimension 2) $Y^{s^t}=[Y_k^t,Y_h^t]'$ of the traffic flowing on the links k, h at time t ($t = 1, \ldots, T$). Let further A^s be the $2 \times N$ sub-matrix of the routing matrix A composed by its k-th and h-th rows.

From the fairly obvious relationship

$$Y^{s^t} = A^s X^t$$

it follows that Y^{s^t} possesses a bivariate normal distribution, with mean vector

$$\mu^s_y = A^s \mu_x$$

and covariance matrix

$$G^s(\theta_x) = A^s \Sigma_x A^{s'}.$$

Hence, the statistical data restricted to the pair $s = (k, h)$ of links

$$Y^s = [Y^{s^1} \, Y^{s^2} \, \ldots \, Y^{s^T}]$$

do have a normal $2T$-variate distribution, with mean vector

$$1_T \otimes (A^s \mu_x)$$

and covariance matrix

$$\Sigma^s = R_T(H) \otimes A^s \Sigma_x A^{s'}$$
$$= R_T(H) \otimes G^s(\theta)$$

where $R_T(H)$ is the $T \times T$ circulant matrix of elements $1, \rho_x(1), \ldots, \rho_x(T - 1)$.

If $l^s_T(\theta_x, H)$ denotes the log-likelihood corresponding to the "partial data" Y^s corresponding to the pair $s = (k, h)$ of links, then the pseudo-log-likelihood can be written as

$$l^p_T(\theta_x, H) = \sum_s l^s_T(\theta_x, H) \qquad (26)$$

where $\sum_s$ is the sum w.r.t. all $\binom{M}{2}$ pairs of links.

The pseudo-likelihood (26) can be further simplifications by approximating each term $l^s(\theta, H)$ as made for the full log-likelihood. Consider the "partial periodogram" for the pair $s = (k, h)$ of links:

$$I^s_T(\omega_j) = w^s_T(\omega_j) w^s_T(\omega_j)'$$

where

$$w_T^s(\omega_j) = \frac{1}{\sqrt{2\pi T}} \sum_{t=1}^{T} Y^{st} e^{it\omega_j}$$

and $\omega_j = (2\pi j)/T - \pi$, $j = 1, \ldots, T$.

Then, using the same reasoning as in Sect. 4.1, the approximation

$$l^s(\theta_x, H) \approx \bar{l}_T^s(\theta, H)$$

$$= -\sum_{j=1}^{T} \{2\log f_x(\omega_j; H) + \log det(G^s(\theta))$$

$$+ f_x(\omega_j; H)^{-1} tr(G^s(\theta_x)^{-1} I_T^s(\omega_j))\}$$

holds. As a consequence, the pseudo-likelihood (26) can be approximated by

$$l_T^p(\theta, H) \approx \bar{l}_T^p(\theta, H) = \sum_s \bar{l}_T^s(\theta, H). \tag{27}$$

The use of the pseudo-likelihood (27) instead of the full likelihood (10) is essentially justified by its computational gain. In fact, each term $l_T^s(\theta, H)$ requires the inversion of a 2×2 matrix. Hence, the complexity of $l_T^p(\theta_x, H)$ is *linear* w.r.t. M.

Generally speaking, the pseudo-likelihood (27) is a special form of composite likelihood (cfr. [31]). A well-known fact (see [31] and references therein) is that composite likelihood methods do not work in the presence of LRD. However, in our case the pseudo-likelihood just factorizes w.r.t. pairs of links, but not w.r.t. time. Hence, it preserves the LRD.

5.2 Properties of the Maximum Pseudo-Likelihood Estimators

The goal of this section is to provide a few asymptotics for the roots of the pseudo-likelihood equation $\bar{l}_T^p(\theta, H) = 0$, denoted in the sequel by $\widehat{\theta}^p$, $\widehat{H}^p$, respectively.

From now on, we will denote by H_0, θ_0 the "true values" of H, θ, respectively.

Proposition 3. *Under the assumptions of Sect. 2, for every θ, H the following result holds:*

$$T^{-1}\bar{l}_T^p(\theta, H) \overset{a.s.}{\to} -Q^p(\theta, H) \tag{28}$$

where

$$Q^{\mathrm{p}}(\theta, H) = Q_1^{\mathrm{p}}(\theta, H) + Q_2^{\mathrm{p}}(\theta, H)$$

$$Q_1^{\mathrm{p}}(\theta, H) = 2 \int_{-\pi}^{\pi} \log f_X(\omega, H)\, d\omega + \frac{2\pi}{\binom{M}{2}} \sum_S^s \log \det G^s(\theta)$$

$$Q_2^{\mathrm{p}}(\theta, H) = \frac{1}{\binom{M}{2}} \sum_S \mathrm{tr}\left(G^s(\theta)^{-1} G^s(\theta_0)\right) \int_{-\pi}^{\pi} \frac{f_X(\omega, H_0)}{f_X(\omega, H)}\, d\omega.$$

Proof. Easy consequence of the law of large numbers for stationary processes. $\square$

Intuitively speaking, as a consequence of Proposition 3, as T increases the roots of equation $\bar{l}_T^{\mathrm{p}}(\theta, H) = 0$ tend to the roots of the equation $Q^{\mathrm{p}}(\theta, H) = 0$. It is not difficult to see that the only roots of such an equation are the true values θ_0, H_0. Under the regularity conditions of Sect. 2, this can be shown using exactly the same arguments as in [7].

Proposition 4. *As T increases, we have:*

$$\widehat{\theta}^{\mathrm{p}} \xrightarrow{a.s.} \theta_0, \quad \widehat{H}^{\mathrm{p}} \xrightarrow{a.s.} H_0$$

$$\sqrt{T} \begin{bmatrix} \widehat{\theta}_x^{\mathrm{p}} - \theta_{x0} \\ \widehat{H}^{\mathrm{p}} - H_0 \end{bmatrix} \xrightarrow{d} \mathcal{N}_{N+1}(0, W^{\mathrm{p}}(\theta_0, H_0)\, V(\theta_0, H_0)\, W^{\mathrm{p}}(\theta_0, H_0)')$$

where $V(\theta_0, H_0)$ is the matrix having elements (23), and $W^{\mathrm{p}}(\theta_0, H_0)$ is a matrix defined exactly as W in Sect. 4.2, except that Q_1, Q_2, Q are replaced by $Q_1^{\mathrm{p}}, Q_2^{\mathrm{p}}, Q^{\mathrm{p}}$, respectively.

Proof. It is sufficient to use the same arguments as in [7]. $\square$

Now, it is not difficult to see that the maximum pseudo-likelihood estimators $\widehat{\theta}^{\mathrm{p}}, \widehat{H}^{\mathrm{p}}$ are asymptotically less efficient than the maximum likelihood estimators $\widehat{\theta}$, $\widehat{H}$. The loss of asymptotic efficiency is the price to pay to reduce computational complexity. In the subsequent section we will compare, through a simulation experiment, the estimators $\widehat{\theta}^{\mathrm{p}}, \widehat{H}^{\mathrm{p}}$ and $\widehat{\theta}, \widehat{H}$.

6 Simulation Study

In this section a simulation study is performed, in order to compare maximum likelihood and pseudo-likelihood estimators (MLE and MPLE, for short) of the S–D traffic intensities.

Fig. 1 Network used in the
simulation study

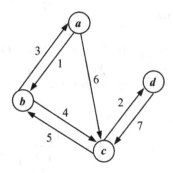

6.1   Simulation Assumptions

The network used in our study is the same already used by Vardi in his seminal
paper [30], as well as in [6, 7]. It is reported in Fig. 1.

The nodes in Fig. 1 are labeled by letters, and the links by numbers. The network
consists of four nodes (hence 12 S–D pairs) and seven unidirectional links.

The assumptions on which our simulation study is based are listed below.

1. The traffic generated satisfies assumptions 1–3 of Sect. 2.
2. All the S–D pairs have the same value of the Hurst parameter.
3. The expected value of traffic follows a Zipf rank-size relationship.
4. The mean–variance relationship for the traffic intensity follows a power law.
5. The coefficient of preference, which determines how the traffic generated by a
   given origin node distributes among all the destinations, is proportional to the
   traffic generated by the destination node.

As remarked in the Introduction, Assumptions 1–2 are well supported in the
literature. In order to clarify the meaning of Assumption 3, suppose the n nodes
are sorted according to the average traffic generated. Denote by μ_{O_j} the expected
traffic generated by jth node, so that $\mu_{O_1} \le \mu_{O_2} \le \cdots \le \mu_{O_n}$. Assumption 4 means
that μ_{O_j}s obey the Zipf law

$$\mu_{O_j} \propto \frac{1}{i^\alpha} \qquad i = 1, \ldots, n. \tag{29}$$

This law, originally formulated in the context of linguistics in [37], is supported by
measurement campaigns performed on the telephone network and on Internet users
(see [22]). Assumption 4 is expressed by the relationship

$$\sigma_{X_i}^2 = \phi \, \mu_{X_i}^c \qquad i = 1, \ldots, N. \tag{30}$$

It was put forward in the paper by [3] and is supported by several measurements
campaigns: [12, 21, 27]. Finally, Assumption 5 is common in tele-traffic studies;

Table 1 Average error over S–D pairs

	Estimation method					
	MLE			MPLE		
	$H = 0.5$	$H = 0.6$	$H = 0.8$	$H = 0.5$	$H = 0.6$	$H = 0.8$
$T = 30$						
$c = 1.0$	13.7	15.4	15.9	14.2	16.4	16.7
$c = 1.5$	15.4	15.5	16.1	16.5	16.8	18.2
$T = 90$						
$c = 1.0$	11.9	12.2	12.8	12.5	12.9	14.7
$c = 1.5$	12.2	12.6	14.7	13.1	13.5	16.9
$T = 120$						
$c = 1.0$	11.6	11.3	12.4	12.2	12.4	14.1
$c = 1.5$	12.1	11.3	14.5	13.0	13.4	16.2

see, *e.g.*, [1]. Consider the ith S–D pair, and denote by O_l its origin node, and by O_m its destination node. The expected traffic intensity for such S–D pair is then

$$\mu_{X_i} = \mu_{O_l} \frac{\mu_{O_m}}{\sum_{k=1}^{n} \mu_{O_k}} \qquad i = 1, \ldots, N. \tag{31}$$

Finally, the long-range-dependent traffic traces are generated by using the Choleski method; see [8].

In our simulation the following set of parameter values have been used.

- $H = 0.5, 0.6, 0.8$.
- Zipf parameter $\alpha = 1$.
- Sample size (traffic traces length) $T = 30, 90, 120$.
- Parameter of the power-law relationship between mean and variance $\phi = 1$ and $c = 1, 1.5$.

6.2 Simulation Results

In this section, the performance of MLE and MPLE of the expected S–D traffic is compared *via* simulation, for different values of the simulation parameters. For each combination of estimation method, value of T (sample size), value of H, the average error over all the S–D pairs (as in [14]) is reported in Table 1.

The estimation method based on the full likelihood clearly outperforms the estimation method based on the pseudo-likelihood. The loss of efficiency of MPLEs is about 15–25 %, if compared to MLEs. However, even for the small-scale network used in the present study, MPLEs do offer a good advantage from a computational point of view. As discussed in the previous section, from a theoretical point of view they have a smaller computational complexity (linear w.r.t M instead of super-quadratic or oven cubic). From a practical point of view, in our simulation experiments the computation time for MLEs ranges from twice to four times the computation time for MPLEs.

References

1. Bear, D.: Principles of Telecommunication Traffic Engineering. IEEE/Peter Peregrinus, London (1988)
2. Bermolen, P., Vaton, S., Juva, I.: Search for optimality in traffic matrix estimation: a rational approach by Cramér-Rao lower bounds. In: Proceedings of the 2nd Conference on Next Generation Internet Design and Engineering (NGI 2006), Valencia (2006)
3. Cao, J., Davis, D., Vander Wiel, S., Yu, B.: Time-varying network tomography: router link data. J. Am. Stat. Assoc. **95**, 1063–1075 (2000)
4. Castro, R., Coates, M., Liang, G., Nowak, R., Yu, B.: Network tomography: recent developments. Stat. Sci. **19**, 499–517 (2004)
5. Conti, P.L., De Giovanni, L., Naldi, M.: Blind maximum-likelihood estimation of traffic matrices in long range dependent traffic. In: Valadas, R., Salvador, P. (eds.) Lectures Notes in Computer Science, pp. 141–154. Springer, Berlin (2009)
6. Conti, P.L., De Giovanni, L., Naldi, M.: Blind maximum likelihood estimation of traffic matrices under long-range dependent traffic. Comput. Networks **54**, 2626–2639 (2010)
7. Conti, P.L., De Giovanni, L., Naldi, M.: Estimation of traffic matrices in the presence of long-memory traffic. Stat. Model. **12**, 29–65 (2012)
8. Conti, P.L., De Giovanni, L., Stoev, S., Taqqu, M.S.: Confidence intervals for the long memory parameter based on wavelets and resampling. Statistica Sinica **18**, 559–579 (2008)
9. Crovella, M.E., Bestavros, A.: Self-similarity in world wide Web taffic: evidence and possible causes. IEEE/ACM Trans. Networking **5**, 835–846 (1997)
10. Fonseca, N., Mayor, G., Neto, C.: On the equivalent bandwidth of self-similar sources. ACM Trans. Model. Comput. Simul. **10**, 104–124 (2000)
11. Goldschmidt, O.: ISP backbone traffic inference methods to support traffic engineering. In: Internet Statistics and Metrics Analysis (ISMA) Workshop, San Diego (Ca) (2000)
12. Gunnar, A., Johansson, M., Telkamp, T.: Traffic matrix estimation on a large IP backbone: a comparison on real data. In: 4th ACM SIGCOMM Conference on Internet Measurement, pp. 149–160, New York (2004) doi: 10.1145/1028788.1028807
13. Hosoya, Y.: A limit theory for long range dependence and statistical inference on related models. Ann. Stat. **25**, 105–137 (1997)
14. Juva, I.: Traffic matrix estimation in the internet: measurement analysis, estimation methods and applications. Ph.D. Thesis in Technology, University of Technology, Helsinki (2007)
15. Juva, I., Vaton, S., Virtamo, J.: Quick traffic matrix estimation based on link count covariances. In: Proceedings of the 2006 IEEE International Conference on Communications (ICC 2006), vol. 2, pp. 603–608, Istanbul (2006)
16. Juva, I., Susitaival, R., Peuhkuri, M., Aalto, S.: Traffic characterization for traffic engineering purposes: analysis of Funet data. In: Proceedings of the 1st EuroNGI Conference on Next Generation Internet Networks (NGI 2005), pp. 404–422, Rome (2005)
17. Juva, I., Susitaival, R., Peuhkuri, M., Aalto, S.: Effects of spatial aggregation on the characteristics of origin-destination pair traffic in Funet. In: Koucheryavy, Y., Harju, J., Sayenko, A. (eds.) Lectures Notes in Computer Science, vol. 4712, pp. 1–12. Springer, Berlin (2007)
18. Leland, W.E., Taqqu, M.S., Willinger, W., Wilson, D.V.: On the self-similar nature of Ethernet traffic (Extended version). IEEE/ACM Trans. Networking **2**, 1–15 (1994)
19. Liang, G., Yu, B.: Estimation in network tomography. IEEE Trans. Signal Process. **51**, 2043–2053 (2003)

20. Magnus, J.R., Neudecker, H.: Matrix Differential Calculus with Applications in Statistics and Econometrics, 3rd edn. Wiley, New York (1999)
21. Medina, A., Taft, N., Salamatian, K., Bhattacharyya, S., Diot, C.: Traffic matrix estimation: existing techniques and new directions. In: Proceedings SIGCOMM02, Pittsburgh (2002)
22. Naldi, M., Salaris, C.: Rank-size distribution of teletraffic and customers over a wide area network. Eur. Trans. Telecomms. **17**, 415–421 (2006)
23. Norros, I., Kilpi, J.: Testing the Gaussian approximation of aggregate traffic. In: Internet Measurement Workshop IMW2002, Marseille (2002)
24. Norros, I., Pruthi, P.: On the applicability of Gaussian traffic models. In: Emstad, P.J., Helvik, B.E., Myskja, A.H. (eds.) The Thirteenth Nordic Teletraffic Seminar, Trondheim (1996)
25. Park, K., Willinger, W.: Self-Similar Network Traffic and Performance Evaluation. Wiley, New York (2000)
26. Park, C., Hernandez-Campos, F., Marron, J.S., Donelson Smith, F.: Long range dependence in a changing Internet traffic mix. Comput. Networks **48**, 401–422 (2005)
27. Susitaival, R., Juva, I., Peuhkuri, M., Aalto, S.: Characteristics of origin-destination pair traffic in Funet. Telecommun. Syst. **33**, 67–88 (2006)
28. Taqqu, M.S., Willinger, W., Sherman, R.: Proof of a fundamental result in selfsimilar traffic modeling. SIGCOMM Comput. Commun. Rev. **27**, 5–23 (1997)
29. Tebaldi, C., West, M.: Bayesian inference on network traffic using link count data. J. Am. Stat. Assoc. **93**, 557–573 (1998)
30. Vardi, Y.: Network tomography: estimating source-destination traffic intensities from link data. J. Am. Stat. Assoc. **91**, 365–377 (1996)
31. Varin, C., Reid, N., Firth, D.: An overview of composite likelihood methods. Statistica Sinica **21**, 5–42 (2011)
32. Vaton, S., Gravey, A.: Iterative Bayesian estimation of network traffic matrices in the case of Bursty flows. In: Internet Measurement Workshop IMW2002, Marseille (2002) http://public. enst-bretagne.fr/~vaton/extended-abstract.ps
33. Veitch, D., Abry, P.: A wavelet based joint estimator of the parameters of long range dependence traffic. IEEE Trans. Inform. Theor. (special issue on "Multiscale Statistical Signal Analysis and its Applications") **45**, 878–897 (1999)
34. Willinger, W., Taqqu, M.S., Leland, W.E., Wilson, D.V.: Self similarity in high-speed traffic: analysis and modelling of Ethernet traffic measurements. Stat. Sci. **10**, 67–85 (1995)
35. Willinger, W., Taqqu, M.S., Leland, W.E., Wilson, D.V.: Self similarity and heavy tails: structural modeling of network traffic. In: Adler, R.J., Feldman, R., Taqqu, M.S. (eds.) A Practical Guide to Heavy Tails: Statistical Techniques and Applications. Birkhäuser, Boston (1998)
36. Yajima, Y.: On estimation of long-memory time series models. Aust. J. Stat. **27**, 303–320 (1985)
37. Zipf, G.K.: Human Behavior and the Principle of Least Effort. Addison-Wesley, New York (1949)

A Newton's Method for Benchmarking Time Series

Tommaso Di Fonzo and Marco Marini

Abstract We present a Newton's method with Hessian modification for benchmarking a time series according to a *growth rates preservation* principle. Unlike the well-known *proportionate first differences* solution by [7], this technique is based on a more natural measure of the movement of the preliminary series, whose dynamic profile is aimed to be preserved as much as possible by the benchmarked series. The computational issues arising from the nonlinearity of the problem can be dealt with by a computationally robust and efficient approach, which results in an effective statistical tool also in a data-production process involving a considerable amount of series.

1 Introduction

The need for benchmarking monthly and quarterly series to annual series arises when time series data for the same target variable are measured at different frequencies with different level of accuracy, and one wishes to remove discrepancies between the annual benchmarks and the corresponding aggregates (either sums or averages) of the sub-annual values. The optimal combination of benchmark levels and short-term movements requires an adjustment which preserves as much as possible the temporal profile of the preliminary infra-annual figures subject to the restrictions provided by the less frequent constraints.

T. Di Fonzo (✉)
Department of Statistical Sciences, University of Padua, Padua, Italy
e-mail: tommaso.difonzo@unipd.it

M. Marini
Statistics Department, International Monetary Fund, Washington, DC, USA
e-mail: mmarini@imf.org

M. Grigoletto et al. (eds.), *Complex Models and Computational Methods in Statistics*,
Contributions to Statistics, DOI 10.1007/978-88-470-2871-5_9,
© Springer-Verlag Italia 2013

The most widely used benchmarking procedure is the modified Denton *Proportionate First Differences* (*PFD*) technique [5, 7]. The *PFD* procedure looks for benchmarked estimates aimed at minimizing the sum of squared proportional differences between the target and the unbenchmarked values, and is characterized by an explicit benchmarking formula involving simple matrix operations.

The *Growth Rates Preservation* (*GRP*) benchmarking procedure by [4] (see also [2, 13]) is based on a more explicit *movement preservation principle*, according to which the sum of squared differences between the growth rates of the target and of the preliminary series is minimized. [1] (p. 100) claims that this is an "ideal" movement preservation principle, "formulated as an explicit preservation of the period-to-period rate of change" of the preliminary series. The *GRP* procedure looks for the solution to a constrained Non Linear Program (*NLP*), according to which $f(\mathbf{x})$, a smooth, non-convex function of the n unknown items of vector $\mathbf{x}$, is minimized subject to m linear equality constraints, $\mathbf{Ax} = \mathbf{b}$, where $\mathbf{A}$ is a known, full row rank $(m \times n)$ matrix, $m < n$, and $\mathbf{b}$ is a known $(m \times 1)$ vector containing the benchmarks.

Both the original algorithm by [4] and a recent proposal by [3] are first-order (i.e., gradient-based) feasible direction methods, which use the Steepest Descent (*SD*) and the nonlinear Conjugate Gradient (*CG*) algorithms, respectively, to solve the above *NLP* problem. However, using only first-derivatives information may result in poorly efficient procedures, characterized by slow convergence and possible troubles in finding actual minima of the objective function.

Still remaining at first-order techniques, more performing unconstrained Quasi-Newton (*QN*) optimization procedures may be considered, which exploit approximate rather than exact second derivatives, provided the original constrained problem be transformed into an unconstrained one. In addition, improvements in both efficiency and robustness may be obtained by considering the true Hessian matrix of the objective function.

In this paper we propose a Newton's method with Hessian modification (*MN*) to calculate *GRP* benchmarked estimates and compare the performance of *MN* with gradient-based procedures (*SD, CG, QN*), in order to show the effectiveness of the proposed benchmarking procedure in terms of both computational efforts and quality of the results.

The paper is organized as follows. In Sect. 2 the *GRP* benchmarking procedure and the way it takes into account a movement preservation principle are discussed, as compared to the classical procedure by [7], described in Sect. 3. An algorithm based on a Newton's method with Hessian modification is described in Sect. 4. In order to analyze the distinctive features of the proposed procedure, in Sect. 5 are presented applications to 61 quarterly series from the EU Quarterly Sector Accounts (EUQSA), and 236 monthly series from the Canadian Monthly Retail Trade Survey (MRTS).

2 Growth Rates Preservation and Temporal Benchmarking

Let b_T, $T = 1, \ldots, m$, and p_t, $t = 1, \ldots, n$, be, respectively, the temporal benchmarks and the high-frequency preliminary values of an unknown target variable x_t. Let s be the aggregation order (e.g., $s = 4$ for quarterly-to-annual aggregation, $s = 12$ for monthly-to-annual aggregation, $s = 3$ for monthly-to-quarterly aggregation), and let $\mathbf{A}$ be a $(m \times n)$ temporal aggregation matrix, converting n high-frequency values into m low-frequency ones (we assume $n = s \cdot m$). If we denote with $\mathbf{x}$ the $(n \times 1)$ vector of high-frequency values, and with $\mathbf{b}$ the $(m \times 1)$ vector of low-frequency values, the aggregation constraints can be expressed as $\mathbf{Ax} = \mathbf{b}$.

Depending on the nature of the involved variables (e.g., flows, averages, stocks), the temporal aggregation matrix $\mathbf{A}$ usually can be written as

$$\mathbf{A} = \mathbf{I}_m \otimes \mathbf{a}^{\mathrm{T}}, \tag{1}$$

where the $(s \times 1)$ vector $\mathbf{a}$ may assume one of the following forms:

1. Flows: $\mathbf{a} = \mathbf{1}_s = (1\ 1\ \ldots\ 1)^{\mathrm{T}}$.
2. Averages: $\mathbf{a} = \frac{1}{s}\mathbf{1}_s$.
3. Stocks (end-of-the-period): $\mathbf{a} = (0\ 0\ \ldots\ 1)^{\mathrm{T}}$.
4. Stocks (beginning-of-the-period): $\mathbf{a} = (1\ 0\ \ldots\ 0)^{\mathrm{T}}$.

Denoting by $\mathbf{p}$ the $(n \times 1)$ vector of preliminary values $(\mathbf{Ap} \neq \mathbf{b})$, we look for a vector of benchmarked estimates $\mathbf{x}^*$ which should be "as close as possible" to the preliminary values, and such that $\mathbf{Ax}^* = \mathbf{b}$.

In an economic time series framework, the preservation of the temporal dynamics (however defined) of the preliminary series is often a major interest of the practitioner. For flows series, [4] considers a *GRP* criterion explicitly related to the growth rate, which is a natural measure of the movement of a time series:

$$f(\mathbf{x}) = \sum_{t=2}^{n} \left(\frac{x_t - x_{t-1}}{x_{t-1}} - \frac{p_t - p_{t-1}}{p_{t-1}} \right)^2 = \sum_{t=2}^{n} \left(\frac{x_t}{x_{t-1}} - \frac{p_t}{p_{t-1}} \right)^2, \tag{2}$$

and look for values x_t^*, $t = 1, \ldots, n$, which minimize the criterion (2) subject to the aggregation constraints $\sum_{t \in T} x_t = b_T$, $T = 1, \ldots, m$. In other words, the benchmarked series is estimated in such a way that its temporal dynamics, as expressed by the growth rates $\dfrac{x_t^* - x_{t-1}^*}{x_{t-1}^*}$, $t = 2, \ldots, n$, be "as close as possible" to the temporal dynamics of the preliminary series, where the "distance" from the preliminary growth rates $\dfrac{p_t - p_{t-1}}{p_{t-1}}$ is given by the sum of the squared differences.

In this paper we consider a more general formulation of the *GRP* benchmarking problem, valid not only for flows variables linked by a simple summation, that is:

$$\min_{\mathbf{x}} f(\mathbf{x}) \qquad \text{subject to} \quad \mathbf{Ax} = \mathbf{b}. \tag{3}$$

3 Modified Denton *PFD*

Reference [7] proposed a benchmarking procedure grounded on the *PFD* between the target and the original series. Reference [5] slightly modified the result of Denton, in order to correctly deal with the starting conditions of the problem. The *PFD* benchmarked estimates are thus obtained as the solution to the constrained quadratic minimization problem

$$\min_{x_t} \sum_{t=2}^{n} \left(\frac{x_t - p_t}{p_t} - \frac{x_{t-1} - p_{t-1}}{p_{t-1}} \right)^2 \quad \text{subject to} \quad \mathbf{Ax} = \mathbf{b}. \tag{4}$$

In matrix notation, the *PFD* benchmarked series is contained in the $(n \times 1)$ vector $\mathbf{x}^{\text{PFD}}$ solution to the linear system [8]

$$\begin{bmatrix} \mathbf{M} & \mathbf{A}^{\text{T}} \\ \mathbf{A} & \mathbf{0} \end{bmatrix} \begin{bmatrix} \mathbf{x}^{\text{PFD}} \\ \lambda \end{bmatrix} = \begin{bmatrix} \mathbf{0} \\ \mathbf{b} \end{bmatrix}, \tag{5}$$

where λ is a $(n \times 1)$ vector of Lagrange multipliers, $\mathbf{M} = \mathbf{P}^{-1}\Delta_n^{\text{T}}\Delta_n\mathbf{P}^{-1}$, $\mathbf{P} = \text{diag}(\mathbf{p})$, and Δ_n is the $((n-1) \times n)$ first differences matrix:

$$\begin{pmatrix} -1 & 1 & 0 & \cdots & 0 & 0 \\ 0 & -1 & 1 & \cdots & 0 & 0 \\ \vdots & \vdots & \vdots & \ddots & \vdots & \vdots \\ 0 & 0 & 0 & \cdots & 1 & 0 \\ 0 & 0 & 0 & \cdots & -1 & 1 \end{pmatrix}.$$

Notice that $\Delta_n^{\text{T}}\Delta_n$ has rank $n-1$, so $\mathbf{M}$ is singular. However, given that matrix $\mathbf{A}$ has full row rank, and provided no preliminary value is equal to zero, the coefficient matrix of system (5) has full rank [8].

Reference [4] use $\mathbf{x}^{\text{PFD}}$ as starting values of the iterative algorithm developed to solve the *NLP* problem (3). This basically depends on two facts:

1. The optimization procedure starts at a feasible point, as $\mathbf{x}^{\text{PFD}}$ clearly is, and at each iteration moves to another feasible point.
2. In the literature [1, 5, 6] it is often claimed that the *PFD* procedure produces results very close to the *GRP* benchmarking, and thus $\mathbf{x}^{\text{PFD}}$, which is considered as a "good" approximation to the *GRP* estimates, is a natural candidate to be used as starting point.

Reference [8] discuss this latter issue, showing that *PFD* and *GRP* benchmarked estimates are close when the variability of the preliminary series and/or its bias are low with respect to the target variable. When this is not the case (e.g., preliminary series with large growth rates and/or bias), the quality of the approximation worsens.

4 Newton's Method with Hessian Modification

Since the criterion (2) is a nonlinear and non-convex function [9], the constrained minimization problem (3) has not linear first-order conditions for a stationary point, and thus it is not possible to find a closed-form solution. On the other hand, provided that both p_t and x_t, $t = 1, \ldots, n - 1$, be different from zero, $f(\mathbf{x})$ is a twice continuously differentiable function, making it possible the use of some iterative minimization algorithms.

In [9] both the gradient vector and the Hessian matrix of function (2) have been analytically derived. In addition, it is shown how the constrained problem (3) in $\mathbf{x}$ can be transformed in the equivalent, reduced unconstrained problem in $\mathbf{x}_Z$:

$$\min_{\mathbf{x}_Z} \tilde{f}(\mathbf{x}_Z), \tag{6}$$

where $\tilde{f}(\mathbf{x}_Z) = f(\bar{\mathbf{x}} + \mathbf{Z}\mathbf{x}_Z)$, $\bar{\mathbf{x}}$ being a feasible $(n \times 1)$ vector (i.e., $\mathbf{A}\bar{\mathbf{x}} = \mathbf{b}$). The $((n - m) \times 1)$ vector $\mathbf{x}_Z$ is such that any feasible point $\mathbf{x}$ can be written as

$$\mathbf{x} = \bar{\mathbf{x}} + \mathbf{Z}\mathbf{x}_Z, \tag{7}$$

where $\mathbf{Z}$ is a $(n \times (n - m))$ basis matrix for the null space of $\mathbf{A}$, whose calculation is numerically efficient and computationally inexpensive (for details, see [9]). A general algorithm for solving the unconstrained minimization problem (6) involves the iterates

$$\mathbf{x}_{Z,k+1} = \mathbf{x}_{Z,k} + \alpha_k \mathbf{d}_k, \quad k = 0, 1, \ldots \tag{8}$$

where α_k is a step-length and $\mathbf{d}_k$ is a direction vector. In its classical form, Newton's method basically consists in determining $\mathbf{d}_k$ in (8) as the solution to the *Newton equations*

$$\left[\nabla^2 \tilde{f}(\mathbf{x}_{Z,k}) \right] \mathbf{d}_k = - \left[\nabla \tilde{f}(\mathbf{x}_{Z,k}) \right]. \tag{9}$$

Since it can fail or diverge, and even if it does converge, it might not converge to a minimum, Newton's method is rarely used in its classical form. A practical version of Newton's method, that is guaranteed to converge and does not assume that $\nabla^2 \tilde{f}(\mathbf{x}_{Z,k})$ is positive definite for all values of k, can be summarized as follows.

1. Specify some initial guess of the solution, $\mathbf{x}_{Z,0}$, and specify a convergence tolerance ϵ.
2. For $k = 0, 1, \ldots$, if $\| \nabla \tilde{f}(\mathbf{x}_{Z,k}) \|_1 < \epsilon$, then stop. Otherwise:

   (a) Compute a modified factorization of the Hessian:

$$\nabla^2 \tilde{f}(\mathbf{x}_{Z,k}) + \mathbf{E} = \mathbf{L}\mathbf{D}\mathbf{L}^\mathsf{T},$$

Table 1 Denton series: iterations, function evaluations and final GRP function

NLP procedure	n. of iterations	n. of function evaluations	Objective function
Steepest descent	36	113	0.04412774
Conjugate gradient	15	57	0.04412700
Quasi-Newton $BFGS$	39	41	0.04411658
Newton with Hessian modification	4	4	0.04411656

where $\mathbf{L}$ and $\mathbf{D}$ are lower triangular and diagonal, respectively, $(n - m) \times (n - m)$ matrices. Then, solve

$$\left(\mathbf{LDL}^{\mathrm{T}}\right)\mathbf{d}_k = -\left[\nabla \tilde{f}\left(\mathbf{x}_{Z,k}\right)\right]$$

for the search direction $\mathbf{d}_k$. Notice that $\mathbf{E}$ will be zero if $\nabla^2 \tilde{f}\left(\mathbf{x}_{Z,k}\right)$ is positive definite.

(b) Perform a line search to determine the new estimate of the solution.

A principal advantage of the Newton's method with Hessian modification is that it converges rapidly when the current estimate of the variables is close to the solution. Its main disadvantage is represented by possible high computational costs, since it requires the derivation, computation, and storage of the Hessian matrix, and the solution of a system of linear equations. This last task could give rise to high computational costs if the dimension of the problem $(n - m)$ is not small and/or the problem is not sparse.

However, for the problem in hand, [9] shows the analytical expressions and the patterns of gradient and Hessian matrix of the problem, so we can take advantage of sparsity, and greatly reduce the computational costs of Newton's method, making it an effective tool in practice.

For example, in the GRP benchmarking of the artificial series of [7], the Newton's method need very few iterations and function evaluations (in both cases, 4) to converge (Table 1), whereas after four iterations three well-known gradient-based algorithms are rather far from the minimum (Fig. 1). However, all the procedures succeed in finding the minimum of the objective function and, according to the quality ranking defined in the next section, all solvers yield "very accurate" solutions, the "best" being given by quasi-Newton and modified Newton's methods.

5   Applications to Quarterly and Monthly Series

In this section we present numerical results about the performance of the Newton's method, as compared to three gradient-based nonlinear minimization methods, in benchmarking 61 quarterly series and 236 monthly series to their annual counterparts. The GRP-benchmarked series are computed by applying

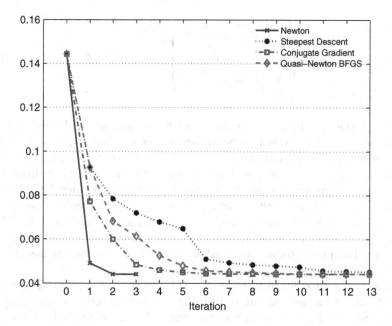

Fig. 1 Denton (1971) series: GRP objective function in the first 14 iterations steps

the following unconstrained nonlinear optimization algorithms to the reduced problem (6):

- Steepest Descent (SD);
- Conjugate Gradient (CG);
- Quasi-Newton Broyden-Fletcher-Goldfarb-Shanno (QN-$BFGS$);
- Newton's method with Hessian modification (MN).

The first dataset consists of 61 raw (not seasonally adjusted) quarterly series from the European Quarterly Sector Accounts (EUQSA), which are not in line with their known annual counterpart. The preliminary series span the period from 1999-Q1 to 2005-Q4 (28 quarters), and annual benchmarks are available for each variable. We consider also 236 monthly seasonally adjusted (SA) series of the Canadian Monthly Retail Trade Survey (MRTS). For 226 out of 236 series, the dataset covers the period from January 1991 to December 2003, while the remaining 10 series start on January 1999. Discrepancies between annually-aggregated SA series and annual benchmarks are usually generated from the application of seasonal adjustment procedures, and benchmarking is often required to restore consistency.

In order to assess the ability of the GRP benchmarked estimates in preserving the dynamics of the preliminary series, as compared to the Denton's PFD solution, we use the two indices [8]:

$$r_q = \left(\frac{\sum_{t=2}^{n} \left| \frac{x_t^{GRP}}{x_{t-1}^{GRP}} - \frac{p_t}{p_{t-1}} \right|^q}{\sum_{t=2}^{n} \left| \frac{x_t^{PFD}}{x_{t-1}^{PFD}} - \frac{p_t}{p_{t-1}} \right|^q} \right)^{\frac{1}{q}}, \qquad q = 1, 2, \tag{10}$$

where the series $\mathbf{x}^{GRP}$ have been calculated using the algorithms outlined above.

Index r_1 can be seen as the ratio between two mean absolute differences between the growth rates of the benchmarked (GRP and PFD, respectively) and the preliminary series. Sometimes this index can be larger than 1, thus indicating that, according to this metric, the movement is better preserved by Denton PFD. When $q = 2$, the index (10) is simply the square root of the ratio between the Causey and Trager movement preservation criteria (2), computed for the GRP and the PFD benchmarked estimates, respectively. Obviously, we expect the GRP technique always reaches a lower (or at least equal) value of the chosen criterion than PFD, and thus the index r_2 should be never larger than 1.

We have used the function minFunc [11] installed on Matlab version 7.7 R2008b [12]. A valuable feature of minFunc is that the scripts of the function are free and available to the user, who can change them according to her/his needs (for details, see [9]).

Convergence is achieved when the norm of the reduced gradient of the objective function is negligible. More precisely, a GRP benchmarked series $\mathbf{x}^* = \bar{\mathbf{x}} + \mathbf{Z}\mathbf{x}_Z^*$ is obtained when

$$\|\nabla \tilde{f}\left(\mathbf{x}_Z^*\right)\|_1 \equiv \sum_{i=1}^{n-m} \left|\tilde{g}_i\left(\mathbf{x}_Z^*\right)\right| \leq 10^{-7}, \tag{11}$$

where $\tilde{g}_i\left(\mathbf{x}_Z^*\right)$, $i = 1, \ldots, n - m$, is the generic element of the reduced gradient vector $\nabla \tilde{f}\left(\mathbf{x}_Z^*\right)$. If condition (11) is not satisfied after 5,000 iterations, the algorithm ends and returns the most recent (feasible) solution.

For comparisons' completeness, we consider also the GRP benchmarked series produced by the DOS-executable programme BMK1.exe, based on the projected steepest descent algorithm by [4], which has been used for a long time by the U.S. Bureau of the Census. We denote this solution with SD-$BMK1$.

The convergence condition of BMK1 is $\dfrac{f\left(\mathbf{x}_{k-1}\right)}{f\left(\mathbf{x}_k\right)} < 1.00001$, which must be fulfilled within 200 iterations. No information on the number of function evaluations is given, and the $\mathbf{x}^{PFD}$ series is returned as the final solution if the algorithm has a breakdown (this never happened for the series we consider in the paper). Due to the limited possibilities of "tuning" the optimization options of BMK1, we used it as a sort of "black-box." The comparisons could thus seem rather unfair. Indeed, in our view such comparisons should only serve to give an idea of the improvements (if any) we can obtain by using the procedures we present in this paper, as compared to the only (as far as we know) currently available public tool for GRP benchmarking.

According to the specialized literature [10], in order to compare different solvers/ algorithms for NLP problems we should consider (1) efficiency, (2) robustness, and (3) quality of solution of the solvers.

Efficiency, which refers to the amount of computation resources needed to find the solution, is generally measured in terms of solver resource time (runtime). Robustness refers to the ability of the solver to succeed in finding one solution and is generally measured by the number of problems for which a feasible solution is produced (the labelling of a solution as either "successfull" or not, is usually summarized by a solve status return code). While considering these two aspects is sufficient when dealing with convex minimization problems (such as in linear programs or for certain quadratic programs), where the found minimum is generally the global one, for non-convex models, which may admit several local minima, other factors involving solution quality play an important role as well. For example, one solver may indeed be more efficient (i.e., faster), but the solution may be worse than that of a solver which is slower in terms of elapsed time.

For the problem in hand, however, robustness is not a concern, since all the techniques we consider are "feasible point methods"—i.e. at each iterate they produce series in line with the temporal aggregation constraints—designed in such a way as they always give solutions "not worse" than $\mathbf{x}^{PFD}$. In other words, in any case a feasible solution, say $\tilde{\mathbf{x}}$, is obtained, such that $A\tilde{\mathbf{x}} = \mathbf{b}$, and $f(\tilde{\mathbf{x}}) \le f(\mathbf{x}^{PFD})$.

Therefore, if we were only interested in the efficiency in finding a local minimum, we would simply look for the fastest solver. Instead, if we wish that the comparison takes into account also the quality of the solution, it seems sensible to consider the best solution within the available ones,

$$\hat{\mathbf{x}} = \arg \min_{\tilde{\mathbf{x}}} f(\tilde{\mathbf{x}}),$$

and to refer to the relative objective value error between $\tilde{\mathbf{x}}$ and $\hat{\mathbf{x}}$. More precisely, given a positive small tolerance δ, the expression

$$\frac{f(\tilde{\mathbf{x}}) - f(\hat{\mathbf{x}})}{f(\hat{\mathbf{x}})} \le \delta \tag{12}$$

can be used to define a simple quality ranking between the solutions provided by different solvers, which turns out to be effective when a large number of problems have to be considered. Clearly, the "true" best objective value corresponds to a choice of $\delta = 0$, but actually a tolerance close to 0 is used (e.g., $\delta = 0.0001$). The threshold values for δ are chosen to discriminate as much as possible the distance between the different solutions. Thus we say that the solution $\tilde{\mathbf{x}}$ is:

1. the best, if expression (12) holds for $\delta = 0.0001$;
2. very accurate, if expression (12) holds for $\delta = 0.001$;
3. accurate, if expression (12) holds for $\delta = 0.01$;
4. acceptable, if expression (12) holds for $\delta = 0.1$.

Table 2 EUQSA and MRTS series: quality of the solutions with different benchmarking procedures

Quality of solution (tol.%)	Denton PFD	SD BMK1	SD	CG	QN BFGS	Newton
	EUQSA (61 series)					
Bad (>10%)	26	1	4	3	1	0
Acceptable (10%)	35	60	57	58	60	61
Accurate (1%)	15	60	49	51	54	61
Very accurate (0.1%)	3	60	30	34	44	61
Best (0.01%)	0	51	21	25	38	61
	MRTS (236 series)					
Bad (>10%)	33	7	6	1	0	0
Acceptable (10%)	203	229	230	235	236	236
Accurate (1%)	96	227	207	216	218	236
Very accurate (0.1%)	3	212	102	125	162	236
Best (0.01%)	0	94	46	74	153	235
	TOTAL (297 series)					
Bad (>10%)	59	8	10	4	1	0
Acceptable (10%)	238	289	287	293	296	297
Accurate (1%)	111	287	256	267	272	297
Very accurate (0.1%)	6	272	132	159	206	297
Best (0.01%)	0	145	67	99	191	296
	TOTAL (%)					
Bad (>10%)	19.87	2.69	3.37	1.35	0.34	0
Acceptable (10%)	80.13	97.31	96.63	98.65	99.66	100
Accurate (1%)	37.37	96.63	86.20	89.90	91.58	100
Very accurate (0.1%)	2.02	91.58	44.44	53.54	69.36	100
Best (0.01%)	0.00	48.82	22.56	33.33	64.31	99.66

In other words, we consider very accurate a solution whose objective function is within 0.1% of the best possible solution, accurate within 1%, and acceptable within 10%. A solution for which the objective value is 10% larger than the best one is considered of bad quality.

Table 2 reports on the quality, according to the metric defined so far, of the solutions found for the 61 EUQSA series and the 236 MRTS series.

The first column refers to the series benchmarked according to the PFD procedure by [7], which is used as starting point by all the NLP solvers considered in this work. Clearly, Denton PFD is not a true GRP benchmarking procedure, but it is generally considered a good approximation of it. In this comparison it is used as a sort of "baseline": for the whole set of 297 series, and with reference to the GRP objective function (2), Denton PFD yields solutions which are acceptable in about 80% of cases, accurate in about 37% and very accurate in about 2%, thus confirming the good approximation property generally claimed in literature. Anyway, in about 20% of cases this does not hold true, as the solutions by Denton PFD attain a GRP criterion which is more than 10% larger than the best one.

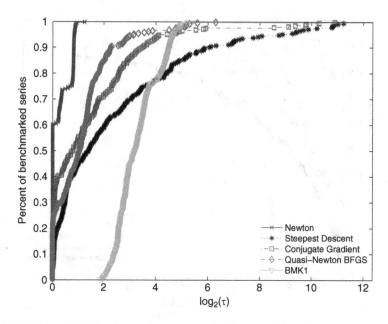

Fig. 2 EUQSA and MRTS series: performance profiles for acceptable solutions

Passing now to consider the "true" NLP solvers, the a priori expectation of a predominance of the Newton's method with Hessian modification is fully confirmed by the results: the Hessian-based procedure never results in solutions of bad quality, and produces by far the best results for almost all series, the unique exception being one of the MRTS series, for which the solution is very accurate, but cannot be considered as the best.

All the gradient-based procedures produce some solutions of bad quality (1 series out of 296 for $QN\text{-}BFGS$, 4 for CG, 8 for $SD\text{-}BMK1$ and 10 for SD). Furthermore, we note that the $SD\text{-}BMK1$ algorithm is uniformly better than the SD solver and produces very accurate solutions in over 91% of cases, a very good performance as compared to more sophisticated optimization algorithms, as CG and $QN\text{-}BFGS$ are.

Figures 2 and 3 show the performance profiles of the considered algorithms (for details, see [9]). The profiles refer to all 297 series and show the performance, as measured by the resource time, when either acceptable solutions or very accurate solutions are considered. The former case gives us information about the efficiency of the solvers, while the latter shows their quality.

The best performance of the Newton's method, in terms of both efficiency and quality, is now confirmed also visually. It is worth noting the performance of $SD\text{-}BMK1$ as compared to the other first-order techniques: for very accurate solutions and large τ on the X-axis, its curve is higher than the other gradient-based methods, due to the capability of producing solutions of better quality.

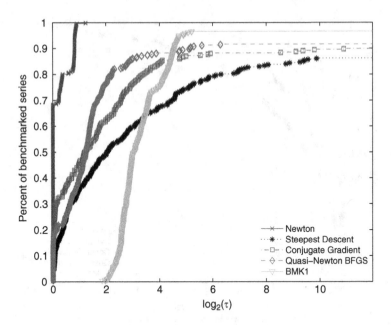

Fig. 3 EUQSA and MRTS series: performance profiles for very accurate solutions

References

1. Bloem, A., Dippelsman, R., Mæhle, N.: Quarterly national accounts manual. Concepts, Data Sources, and Compilation. International Monetary Fund, Washington DC (2001)
2. Bozik, J.E., Otto, M.C.: Benchmarking: evaluating methods that preserve month-to-month changes. Bureau of the Census – Statistical Research Division, RR-88/07 (1988), http://www.census.gov/srd/papers/pdf/rr88-07.pdf
3. Brown, I.: An empirical comparison of constrained optimization methods for benchmarking economic time series. In: JSM Proceedings. Business and Economic Statistics Section. American Statistical Association, Alexandria (2010)
4. Causey, B., Trager, M.L.: Derivation of Solution to the Benchmarking Problem: Trend Revision. Unpublished research notes, U.S. Census Bureau, Washington D.C. (1981) Available as an appendix in Bozik and Otto (1988)
5. Cholette, P.A.: Adjusting sub-annual series to yearly benchmarks. Survey Methodol. **10**, 35–49 (1984)
6. Dagum, E.B., Cholette, P.A.: Benchmarking, Temporal Distribution, and Reconciliation Methods for Time Series. Springer, New York (2006)
7. Denton, F.T.: Adjustment of monthly or quarterly series to annual totals: an approach based on quadratic minimization. JASA **333**, 99–102 (1971)
8. Di Fonzo, T., Marini, M.: Benchmarking and movement preservation. Evidences from real-life and simulated series. Working Paper n. 14/2010, Department of Statistical Sciences, University of Padua (2010), http://www.stat.unipd.it/ricerca/fulltext?wp=422
9. Di Fonzo, T., Marini, M.: A Newton's method for benchmarking time series according to a growth rates preservation principle. Working Paper n. 07/2011, Department of Statistical Sciences, University of Padua (2011), http://www.stat.unipd.it/ricerca/fulltext?wp=432

10. Mittelmann, H.D., Pruessner, A.: A server for automated performance analysis of benchmarking data. Optim. Methods Software **21**, 105–120 (2006)
11. Schmidt, M.: The minFunc Toolbox for Matlab. March 15, 2011. http://www.cs.ubc.ca/~schmidtm (2006)
12. The MathWorks: Optimization Toolbox™4 User's Guide. Natick, MA (2009)
13. Trager, M.L.: Derivation of Solution to the Benchmarking Problem: Relative Revision. Unpublished research notes, U.S. Census Bureau, Washington D.C., (1982) Available as an appendix in Bozik and Otto (1988)

Spatial Smoothing for Data Distributed over Non-planar Domains

Bree Ettinger, Tiziano Passerini, Simona Perotto, and Laura M. Sangalli

Abstract We consider the problem of surface estimation and spatial smoothing over non-planar domains. In particular, we deal with the case where the data or signals occur on a domain that is a surface in a three-dimensional space. The application driving our research is the modeling of hemodynamic data, such as the shear stress and the pressure exerted by the blood flow on the wall of a carotid artery. The regression model we propose consists of two key phases. First, we conformally map the surface domain to a region in the plane. Then, we apply existing regression methods for planar domains, suitably modified to take into account the geometry of the original surface domain.

1 Introduction

In this paper, we deal with data that are observed over non-planar bi-dimensional domains. The motivating application is modeling the shear stress generated by the blood flow over the wall of an internal carotid artery affected by an aneurysm. For instance, Fig. 1 shows the geometry of a possible surface domain of interest: the observed values of the wall shear stress are shown by a color map over the domain itself. This type of data structure, where the quantity of interest is referred to a non-planar domain, occurs in a number of different applications. Another fascinating application in the medical field is, e.g., the study of hemodynamic signals over the cortical surface. Environmental and geostatistical sciences also offer several applications with these types of data structures.

B. Ettinger · S. Perotto · L.M. Sangalli (✉)
MOX – Department of Mathematics, Politecnico di Milano, Milan, Italy
e-mail: bree.ettinger@polimi.it; simona.perotto@polimi.it; laura.sangalli@polimi.it

T. Passerini
Math & Science Center, Emory University, Atlanta, GA
e-mail: tiziano@mathcs.emory.edu

M. Grigoletto et al. (eds.), *Complex Models and Computational Methods in Statistics*,
Contributions to Statistics, DOI 10.1007/978-88-470-2871-5_10,
© Springer-Verlag Italia 2013

Fig. 1 Wall shear stress
modulus at the systolic peak
on a real internal carotid
artery geometry affected by
an aneurysm (data from the
AneuRisk project, http://mox.
polimi.it/it/progetti/aneurisk/)

Unfortunately, few methods are available for smoothing data over non-planar domains (specifically, over bi-dimensional Riemannian manifolds): we recall the nearest neighbor averaging method (see, e.g., [7]) and the sophisticated heat kernel smoothing method proposed in [4]. Here, we adopt a Functional Data Analysis approach and propose a regression method that efficiently handles these data structures. The proposed method consists of two steps: first we map the original surface domain to a flat domain, and then, we properly modify existing spatial regression methods suited to deal with data on planar domains. In particular, to flatten the original surface domain we use a conformal map. The main advantage of using of a conformal map, with respect to any other map, is that it preserves the angles of the original surface domain in the planar domain. The spatial regression method we use is the penalized least square estimation technique proposed in [17] and later generalized in [19]. In our proposed method, the penalty is modified with a contribution from the conformal flattening map, describing the corresponding deformation of the domain.

The paper is organized as follows. Section 2 describes the motivating applied problem in more detail. In Sect. 3, we first recall the spatial regression methods defined in [17] and then introduce the new approach. Section 4.1 provides a simple simulation study. In Sect. 4.2 we apply the proposed method to the study of hemodynamic data. Finally, in Sect. 5 we discuss possible extensions and future directions for the proposed approach.

2 Motivating Applied Problem

The research described in this paper is motivated by the analysis of data within the AneuRisk Project, a scientific endeavor that investigates the pathogenesis of cerebral aneurysms, in an interdisciplinary effort combining the experience of practitioners from neurosurgery and neuroradiology with that of researchers from statistics, numerical analysis, and bio-engineering. For a description of the AneuRisk Project, we refer the interested reader to the website http://mox.polimi.it/it/progetti/aneurisk and references therein.

Cerebral aneurysms are deformations of cerebral vessels characterized by a bulge of the vessel wall. Figure 1 shows an example of an internal carotid artery, one of the main arteries bringing blood to the brain, affected by an aneurysm. The origin of aneurysms is considered to be the result of a complex interplay among systemic effects, biomechanical properties of the vessel wall and the continuous effect of the forces exerted by the blood flow on the vessels. These hemodynamic forces depend on the vessel morphology itself. The study of these interactions and their role on the pathogenesis of aneurysms has been the main goal of the AneuRisk Project. The first studies available in literature on the pathology of aneurysms restrict their attention to the aneurysm sac. In contrast, the AneuRisk Project has investigated the morphological and hemodynamic features of the parent vasculature, i.e., the vessel hosting the aneurysm and the upstream vasculature, with the goal of highlighting possible causes of aneurysm onset, development, and rupture (see [10, 20]).

In this paper, we analyze hemodynamic data on the real anatomy of internal carotid arteries. The internal carotid artery geometry is reconstructed from three-dimensional angiographic images, belonging to the AneuRisk data warehouse; for details on vessel geometry reconstruction see, e.g., [11]. The hemodynamic quantities, wall shear stress and pressure, have been simulated in [9] via Computational Fluid Dynamics. As detailed in [9, 10], the blood has been modeled as a Newtonian fluid, and its dynamics has been described by means of the incompressible Navier–Stokes equations. The geometry of the carotid artery has been assumed to be fixed in time, since compliance effects are expected to be negligible in this vascular district. Proper boundary conditions have been devised to ensure that the flow regime is comparable among all the simulated cases. For each case, blood velocity and blood pressure have been simulated over three heart beats, and the wall shear stress has been computed in a post-processing step. Figure 1 shows the simulated wall shear stress modulus at the systolic peak on a real three-dimensional geometry. The hemodynamic data are referred to points (x_1, x_2, x_3) on the artery wall, a bi-dimensional non-planar domain. In [3, 19] some first analyses of these data were performed, by flattening a simplified version of the carotid domain. In particular, a new coordinate system is defined by (s, r, θ), where s is the curvilinear abscissa along the artery centerline, r the artery radius, and θ the angle of the surface point with respect to the artery centerline. The domain is then reduced to the plane $(s, \theta * \bar{r})$, where $\bar{r}$ is the average carotid radius on the carotid tract considered. This planar rectangular domain is essentially obtained by cutting the artery wall along the axial direction given by s and then opening and flattening the artery wall. This planar domain is equivalent to a simplified three-dimensional artery geometry, where the radius is kept fixed to a constant value and the curvature of the artery is not taken into account. The map just described will be referred to in the following as the *angular map*. Existing spatial regression methods for the planar setting have been applied to the flattened simplified carotid geometry; in particular [19] employs Spatial Spline Regression (SSR) models. Notice that, by flattening the domain with the angular map and then applying a spatial regression method for planar domains, any information related to the vessel radius and curvature is lost; even though these two geometrical quantities greatly influence the hemodynamics in the artery and statistically discriminate aneurysm presence and location (see, e.g., [20]). Moreover, to have a bijective angular map, it is necessary to exclude the aneurysmal sac (otherwise, different points on the carotid wall would be mapped to the same point on the plane).

In the next section, we recall the SSR model for planar domains used in [19] and then introduce the SSR model for non-planar domains. This model allows us to consider the carotid geometry in its actual complexity, including the varying radius and curvature, and without any need to remove of the aneurysmal sac.

3   Spatial Spline Regression Models

3.1   Spatial Spline Regression Model for Planar Domains

In this section, we present the SSR models for planar domains introduced in [17] and the generalized version provided in [19] (see also [15, 16]).

Let $\{\mathbf{u}_i = (u_i, v_i); \ i = 1, \ldots, n\}$ be a set of n fixed data locations on a bounded regular domain $\Omega \subset \mathbb{R}^2$. Let z_i be the real-valued variable of interest observed at point $\mathbf{u}_i$. Assume the model

$$z_i = f(\mathbf{u}_i) + \epsilon_i \qquad\qquad i = 1, \ldots, n \qquad\qquad (1)$$

where ϵ_i are independent observational errors with zero mean and constant variance, and f is a twice continuously differentiable real-valued function to be estimated. According to the SSR model, the estimate of f is found by minimizing the following functional

$$\sum_{i=1}^{n}(z_i - f(\mathbf{u}_i))^2 + \lambda \int_{\Omega} (\Delta f)^2 \mathrm{d}\Omega, \qquad\qquad (2)$$

i.e., a sum of squared errors regularized via the L^2-norm of the Laplacian of f

$$\Delta f = \frac{\partial^2 f}{\partial u^2} + \frac{\partial^2 f}{\partial v^2}.$$

The Laplacian of f measures the local curvature of f. Hence in (2) via the penalty we are essentially controlling the roughness of the solution. Moreover, the Laplacian is invariant with respect to Euclidean transformations of the domain and this ensures that the smoothness of the estimate does not depend on the arbitrarily chosen coordinate system.

The estimation problem (2) cannot be solved analytically. An approximate solution is found by resorting to a finite element approach. The finite element method is largely employed to approximate partial differential equations and it is widely used in engineering applications (for an introduction to the finite element framework, see, e.g., [13]). The strategy is very similar to univariate splines. The

Fig. 2 Three-dimensional triangular mesh approximating a non-planar test domain

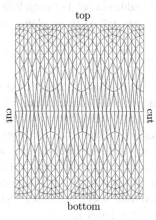

Fig. 3 The planar triangulation obtained by conformally flattening the domain in Fig. 2

top

cut | cut

bottom

Fig. 4 The planar triangulated domain obtained by the angular flattening of the domain in Fig. 2

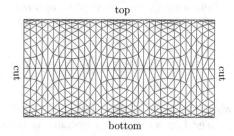

top

cut | cut

bottom

finite element approach subdivides the domain into small disjoint elements and then it yields a local polynomial function on each of these elements, in such a way that the union of all these functions is continuous and closely approximates the solution. This simplified problem becomes computationally tractable thanks to the suitable choice of the basis functions for the space of piecewise polynomials. Convenient domain partitions are provided by triangular meshes (see Figs. 2–4 for some examples). A basis of piecewise polynomials is thus considered over a triangulation of the domain, the simplest being the one spanning the space of all the continuous functions which are linear when restricted to any triangle of the mesh. Thanks to the intrinsic construction of the finite element space, solving the estimation problem (2) reduces to solving a linear system. In particular, the estimator of f turns out to be linear with respect to the observed data values, so that classical inferential tools may be readily derived (see [19]).

3.2   Spatial Spline Regression Model for Non-planar Domains

Now, we consider the problem where the n fixed data locations $\{\mathbf{x}_i = (x_{1i}, x_{2i}, x_{3i}); \ i = 1, \ldots, n\}$ lie over a non-planar domain Σ, where Σ is a surface embedded in $\mathbb{R}^3$. For each location $\mathbf{x}_i$, a real-valued random variable of interest, z_i, is observed. As in the planar case, we assume the model

$$z_i = f(\mathbf{x}_i) + \epsilon_i \qquad\qquad i = 1, \ldots, n \qquad (3)$$

where ϵ_i are independent observational errors with zero mean and constant variance, and f is a twice continuously differentiable real-valued function defined on the surface domain Σ; our aim is to estimate this function. We highlight that our purpose here is far from the one proposed, for instance in [2, 12] and references therein, where statistical tools for data belonging to manifolds are developed. In our case the manifold is just the support of data, in the sense that the data are referred to locations lying on the manifold. We do not have any interest in analyzing the properties of the manifold itself, but rather use its geometrical properties when dealing with data occurring over it.

Following (2), we propose to estimate f in (3) by minimizing the following penalized sum of squared error functional

$$J_\lambda(f(\mathbf{x})) = \sum_{i=1}^{n} (z_i - f(\mathbf{x}_i))^2 + \lambda \int_\Sigma (\Delta_\Sigma f(\mathbf{x}))^2 \, d\Sigma, \qquad (4)$$

where Δ_Σ is the Laplace–Beltrami operator for functions defined over the surface Σ. The Laplace–Beltrami operator is indeed the generalization of the common Laplacian: it can be used to operate on functions defined on surfaces in Euclidean spaces (see, e.g., [5]). Note that we use the convention that Δ_Σ denotes the Laplace–Beltrami operator on the surface Σ, while Δ denotes the standard Laplacian on a planar domain.

In [6], we show that it is possible to solve the estimation problem (4) by exploiting existing techniques for planar domains. In particular, we propose reducing (4) to a problem over a planar domain. To do this, we flatten Σ by means of a conformal map. Specifically, for the surface domain Σ we define a map X such that

$$X : \Omega \to \Sigma$$
$$\mathbf{u} = (u, v) \mapsto \mathbf{x} = (x_1, x_2, x_3) \qquad (5)$$

where Ω is an open, convex, and bounded set in $\mathbb{R}^2$. Denote by $X_u(\mathbf{u})$ and $X_v(\mathbf{u})$ the column vectors of first order partial derivatives of X with respect to u and v, respectively. For the map X to be conformal, we require $\|X_u(\mathbf{u})\| = \|X_v(\mathbf{u})\|$ and $\langle X_u(\mathbf{u}), X_v(\mathbf{u}) \rangle = 0$, for any $\mathbf{u} \in \Omega$, where $\langle \cdot, \cdot \rangle$ denotes the standard Euclidean scalar product and $\| \cdot \|$ is the corresponding norm. Let us also define the (space-dependent) metric tensor as the following symmetric positive definite matrix

$$G(\mathbf{u}) := \begin{pmatrix} \|X_u(\mathbf{u})\|^2 & \langle X_u(\mathbf{u}), X_v(\mathbf{u}) \rangle \\ \langle X_v(\mathbf{u}), X_u(\mathbf{u}) \rangle & \|X_v(\mathbf{u})\|^2 \end{pmatrix} = \begin{pmatrix} g_{11}(\mathbf{u}) & g_{12}(\mathbf{u}) \\ g_{21}(\mathbf{u}) & g_{22}(\mathbf{u}) \end{pmatrix}.$$

Set $W(\mathbf{u}) := \sqrt{\det(G(\mathbf{u}))}$, and denote by $G^{-1}(\mathbf{u}) = \{g^{ij}(\mathbf{u})\}_{i,j=1,2}$ the inverse of the matrix $G(\mathbf{u})$. Using this notation, for a function $f \circ X \in \mathcal{C}^2(\Omega)$, the Laplace–Beltrami operator associated with the surface Σ can be expressed as

$$\Delta_\Sigma f(\mathbf{x}) = \frac{1}{W(\mathbf{u})} \sum_{i,j=1}^{2} \partial_i (g^{ij}(\mathbf{u}) W(\mathbf{u}) \partial_j f(X(\mathbf{u})))$$

where $\mathbf{u} = X^{-1}(\mathbf{x})$. In [6], we show that (4) can be equivalently expressed as the following problem over the planar domain Ω:

$$J_\lambda(f(X(\mathbf{u}))) \tag{6}$$

$$= \sum_{i=1}^{n} (z_i - f(X(\mathbf{u}_i)))^2 + \lambda \int_\Omega \left[\frac{1}{W(\mathbf{u})} \sum_{i,j=1}^{2} \partial_i (g^{ij}(\mathbf{u}) W(\mathbf{u}) \partial_j f(X(\mathbf{u}))) \right]^2 W(\mathbf{u}) d\Omega$$

where $X(\mathbf{u}_i) = \mathbf{x}_i$. Moreover, for conformal coordinates, the functional J_λ reduces to

$$J_\lambda(f(X(\mathbf{u}))) = \sum_{i=1}^{n} (z_i - f(X(\mathbf{u}_i)))^2 + \lambda \int_\Omega \left[\frac{1}{\sqrt{W(\mathbf{u})}} \Delta f(X(\mathbf{u})) \right]^2 d\Omega \tag{7}$$

where Δf is the standard Laplacian over the planar domain Ω. Therefore, this problem turns out to be a modification of the estimation problem presented in Sect. 3.1.

From a computational viewpoint, the conformal map in (5) may be approximated via finite elements. The planar finite elements mentioned in Sect. 3.1 can be adapted to a three-dimensional triangular mesh. In [8] a technique based on finite elements is specifically developed for flattening tubular surfaces (in particular, a portion of the colon). We resort to a similar approach since the wall of the carotid artery is indeed a tubular surface. This approach to estimating the conformal map uses a three-dimensional triangular mesh that approximates the original surface domain Σ. The three-dimensional mesh is flattened into a planar triangular mesh that discretizes Ω via the finite element approximation to the conformal map. One benefit of using a conformal map is that it preserves angles and thus shapes, i.e., the original triangulation.

Figures 2–4 illustrate the flattening of a test surface domain. Figure 2 shows an original non-planar domain approximated by a three-dimensional triangular mesh. Figure 3 displays the conformally equivalent planar triangulated domain. In contrast, Fig. 4 shows the planar domain obtained with the angular map described in Sect. 2. The sides of the planar triangulations are labeled to have a correspondence with respect to the surface in Fig. 2. In particular, the sides of the planar triangulation labeled with "bottom" and "top" correspond to the bottom and to the top open boundaries of the original three-dimensional domain. The two sides indicated by "cut" correspond to a cut along the three-dimensional domain, connecting the two open boundaries of the surface, that is introduced when calculating the flattening map (see [8]).

After the conformal flattening, we are ready to apply the estimation method in Sect. 3.1 and described in detail in [19], with the variant provided in (7), to accommodate for the domain deformation implied by the flattening phase. Note also that the estimates along the two "cut" sides have to coincide; this is in fact an artificial cut. To prevent a seam, we have to take care to maintain the periodicity of the estimate along the "cut" edges (see [3, 6, 19]). Similar to SSR over planar domains, the estimator of f is linear with respect to the observed data values, so that classical inferential tools may be derived. In fact, many of the properties of SSR over planar domains hold for SSR over non-planar domains as we demonstrate in [6].

4 Simulations and Applications to Real Data

4.1 Simulations Studies

In this section, we provide the results of a first simulation study, illustrating the performance of the proposed smoothing technique over non-planar domains. In particular, we compare the results obtained via the proposed SSR model for non-planar domains with those yielded by the SSR model for planar domains combined with a simple angular flattening (see Sect. 2). Notice that the methods differ in two ways. The first is the flattening map. For the SSR model over non-planar domains, we have a triangulation which preserves the shapes of the triangles in the original mesh since it is generated by a conformal map. Instead, the triangulation generated by the angular map does not preserve the shape of the triangles in the original mesh. The second difference is the penalty. SSR models over non-planar domains use information from the conformal flattening map to adjust for the domain deformation implied by the map, hence considering the full three-dimensional domain. SSR models over planar domain does not utilize any information from the flatting and thus has no memory of the geometry in the original three-dimensional domain.

Fig. 5 Three test surface domains. On each surface, the color map indicates one of the selected test functions $f(x_1, x_2, x_3) = a_1 \sin(2\pi x_1) + a_2 \sin(2\pi x_2) + a_3 \sin(2\pi x_3) + 1$, with coefficients a_1, a_2, and a_3 randomly generated from independent normal distributions with mean one and standard deviation one

Fig. 6 On each test surface, at each of the data location $\mathbf{x}_i$, coinciding with the nodes of the three-dimensional meshes approximating the surface domains, independent normally distributed errors with mean zero and a standard deviation 0.5 are added to the test function; the color maps are obtained by linear interpolation of the resulting noisy observations

For these simulations, three domains, approximated by three-dimensional triangular meshes, are considered (see Fig. 5). Each geometry is topologically equivalent to a cylinder. Over each of these non-planar domains, we consider 50 test functions, having the form $f(x_1, x_2, x_3) = a_1 \sin(2\pi x_1) + a_2 \sin(2\pi x_2) + a_3 \sin(2\pi x_3) + 1$ with coefficients a_i, for $i = 1, 2, 3$, randomly generated from independent normal distributions with mean one and standard deviation one. The data locations $\mathbf{x}_i$ coincide with the nodes of the three-dimensional meshes. The noisy observations z_i in correspondence with the locations $\mathbf{x}_i$, for $i = 1, \ldots, n$, are obtained by adding independent normally distributed errors, with mean zero and a standard deviation 0.5, to the test function, in accordance with the model (3). An example of a test function and the corresponding level of noise is illustrated on each geometry in Figs. 5 and 6, respectively.

For each simulation replicate, optimal values of the smoothing parameter λ in (2) and (4) are selected by generalized cross validation for both the models on planar and non-planar domains, as described in [6, 19], respectively.

Table 1 shows the median and inter-quantile ranges of the Mean Square Errors (MSE) of f estimators over the 50 simulations. The table also reports the results of pairwise Wilcoxon tests verifying if the distribution of MSE for the

Table 1 Median (inter-quantile ranges) of MSE of f estimators over the 50 simulations; p-values of pairwise Wilcoxon tests verifying if the distribution of MSE for the estimates provided by SSR over non-planar domains is stochastically lower than the distribution of the MSE for the estimates provided by SSR method over planar domains

MSE	Geometry 1	Geometry 2	Geometry 3
angular map + SSR over planar domains	0.027 (0.018)	0.127 (0.130)	0.111 (0.153)
SSR over non-planar domains	0.025 (0.017)	0.104 (0.095)	0.068 (0.055)
SSR over non-planar vs. SSR over planar	0.016	$5.3e^{-10}$	$3.7e^{-9}$

Fig. 7 The estimates provided by SSR over non-planar domains, with values of λ selected by generalized cross-validation

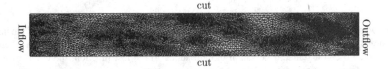

Fig. 8 Planar triangulation generated via the conformal flattening of the mesh approximating the wall of the internal carotid artery in Fig. 9. The sides of the planar triangulation are labeled to correspond with Fig. 9. In particular, the sides of the planar triangulation labeled with "Inflow" and "Outflow" correspond to the open ends of the carotid artery. The sides indicated by "cut" correspond to a longitudinal cut along the artery wall, connecting the open boundaries of the artery

estimates provided by SSR over non-planar domains is stochastically lower than the distribution of the MSE for the estimates provided by SSR method over planar domains. The p-values of these tests show that the MSE of SSR over non-planar domains estimates are significantly lower than the ones of SSR over planar domains, uniformly over the three surface domains considered. Figure 7 displays the estimates provided by SSR over non-planar domains for the three test functions with added noise shown in Fig. 6.

4.2 Application to Hemodynamic Data

This section applies the proposed smoothing technique over non-planar domains to the modeling of the hemodynamic data described in Sect. 2. Figure 8 displays the

Fig. 9 Estimate of wall shear stress modulus at the systolic peak obtained with SSR over non-planar domains with smoothing parameter $\lambda = 0.1$

Outflow

Inflow

planar triangulated domain obtained from the three-dimensional triangulated artery wall, via the computation of the conformal map. Notice the area close to the "Outflow" side where the flattened mesh is very fine; this corresponds to the aneurysmal sac. Recall that the aneurysmal sac has to be removed from the domain when using the simpler angular map. Figure 9 shows the estimate of wall shear stress modulus obtained with SSR over non-planar domains with smoothing parameter $\lambda = 0.1$.

The obtained patient-specific estimates will be used for statistical analyses across patients. These analyses aim to detect recurrent hemodynamic patterns, common across patients, and relate them to presence and location of the pathology, and to rupture risk; furthering the investigation of the origin and pathogenesis of aneurysms. Notice that these analyses also requires appropriate registration of the patient-specific internal carotid artery geometries (see, e.g., [3]).

5   Conclusions and Future Developments

The goal pursued in this paper is to check the capabilities of SSR over non-planar domains. The simple simulation study reported here provides the first evidence of the good properties of the proposed model. In effect, showing that SSR model over non-planar domains provides better estimates than those obtained first by flattening the domain via an angular map and then applying SSR models for planar domains over the flattened domain without accounting for the domain deformation.

Within the framework of the proposed SSR model over non-planar domains, it is also possible to include spatially distributed covariates, as in [19]. In the application to hemodynamics data, this, for instance, would allow the inclusion of the values of blood pressure observed over the artery wall; the pressure could thus be used as a control variable, studying also the relationship between pressure and wall-shear stress, and evaluating how this affects aneurysm pathogenesis. The proposed model can also be extended to data in higher dimensions. In particular, we plan to generalize the model to the case of functional response variables instead of the scalar responses considered in this paper.

Another challenging application for the proposed model is, e.g., the identification of areas of activation for hemodynamics signals over a cortical surface. The cortical surface is a sophisticated geometry that serves as the domain of the signal. A finite element method for conformally flattening the cortical surface in shown in [1];

the proposed SSR model for non-planar domains could thus be used also for this application.

The proposed models have been implemented in R [14] and Matlab. Both code versions, fully integrated with the fda packages in R [18] and Matlab, shall be released shortly.

Acknowledgments This work was funded by MIUR Ministero dell'Istruzione dell'Università e della Ricerca, *FIRB Futuro in Ricerca* research project "Advanced statistical and numerical methods for the analysis of high dimensional functional data in life sciences and engineering" (see http://mox.polimi.it/~sangalli/firb.html), and by the program Dote Ricercatore Politecnico di Milano—Regione Lombardia, research project "Functional data analysis for life sciences." We are grateful to Franco Dassi for help with the 3D meshes and to Piercesare Secchi for interesting discussions.

References

1. Angenent, S., Haker, S., Tannenbaum, A., Kikinis R.: On the Laplace–Beltrami operator and brain surface flattening. IEEE Trans. Med. Imag. **18**, 700–701 (1999)
2. Bhattacharya, A., Bhattacharya, R.N.: Nonparametric statistics on manifolds with applications to shape spaces. Pushing the Limits of Contemporary Statistics: Contributions in Honor of Jayanta K. Ghosh, IMS Collections, vol. 3, pp. 282–301 (2008)
3. Boneschi, A.: Functional data analysis of cfd simulations: the systolic wall shear stress map of the internal carotid artery. Master Thesis. Dipartimento di Matematica "F.Brioschi," Politecnico di Milano, Italy (2010). Available at http://mox.polimi.it/it/progetti/pubblicazioni/tesi/boneschi.pdf
4. Chung, M.K., Robbins, S.M., Dalton, K.M., Davidson, R.J., Alexander, A.L., Evans, A.C.: Cortical thickness analysis in autism with heat kernel smoothing. NeuroImage **25**, 1256–1265 (2005)
5. Dierkes, U., Hildebrant, S., Küster, A., Wohlrab, O.: Minimal Surfaces (I). Springer, Berlin (1992)
6. Ettinger, B., Perotto, S., Sangalli, L.M.: Spatial regression models over bidimensional manifolds. Tech.rep.MOX, Dipartimento di Matematica, Politecnico di Milano. Submitted (2012)
7. Hagler, D.J., Saygin, A.P., Sereno, M.I.: Smoothing and cluster thresholding for cortical surface-based group analysis of fMRI data. NeuroImage **33**, 1093–1103 (2006)
8. Haker, S., Angenent, S., Tannenbaum, A., Kikinis, R.: Nondistorting flattening maps and the 3-D visualization of colon CT images. IEEE Trans. Med. Imag. **19**, 665–670 (2000)
9. Passerini, T.: Computational hemodynamics of the cerebral circulation: multiscale modeling from the circle of Willis to cerebral aneurysms. PhD Thesis. Dipartimento di Matematica, Politecnico di Milano, Italy (2009). Available at http://mathcs.emory.edu/
10. Passerini, T., Sangalli, L.M., Vantini, S., Piccinelli, M., Bacigaluppi, S., Antiga, L., Boccardi, E., Secchi, P., Veneziani, V.: An integrated statistical investigation of internal Carotid arteries of patients affected by Cerebral Aneurysms. Cardiovasc. Eng. Technol. **3**, 26–40 (2012)
11. Piccinelli, M., Veneziani, A., Steinman, D.A., Remuzzi, A., Antiga, L.: A framework for geometric analysis of 852 vascular structures: applications to cerebral aneurysms. IEEE Trans. Med. Imag. **28**(8), 1141–1155 (2009)
12. Pennec, X.: Intrinsic statistics on Riemannian manifolds: basic tools for geometric measurements. J. Math. Imag. Vision **25**, 127–154 (2006)
13. Quarteroni, A.: Numerical models for differential problems. vol. 2 MS&A. Modeling, Simulation and Applications. Springer-Verlag Italia, Milano (2009)

14. R Development Core Team: R: A Language and Environment for Statistical Computing. R Foundation for Statistical Computing, Vienna, Austria (2010), http://www.R-project.org
15. Ramsay, J.O., Ramsay, T.O., Sangalli, L.M.: Spatial functional data analysis. In: Recent Advances in Functional Data Analysis and Related Topics, Contributions to Statistics, pp. 269–276. Physica-Verlag Springer, Wurzburg (2011a)
16. Ramsay, J.O., Ramsay, T.O., Sangalli, L.M.: Spatial spline regression models for data distributed over irregularly shaped regions. In: Proceedings of S.Co.2011 Conference. Available at http://sco2011.stat.unipd.it (2011b)
17. Ramsay, T.: Spline smoothing over difficult regions. J. R. Stat. Soc. Ser. B Stat. Methodol. **64**, 307–319 (2002)
18. Ramsay, J.O., Wickham, H., Graves, S., Hooker, G.: fda: Functional Data Analysis. R package version 2.2.5, http://CRAN.R-project.org/package=fda (2010)
19. Sangalli, L.M., Ramsay, J.O., and Ramsay, T.O.: Spatial spline regression models, J. R. Stat. Soc. Ser. B Stat. Methodol. Part 4, **75**, 1–23 (2013)
20. Sangalli, L.M., Secchi, P., Vantini, S., Veneziani, A.: A case study in exploratory functional data analysis: geometrical features of the internal carotid artery. J. Am. Statist. Assoc. **104**, 37–48 (2009)

Volatility Swings in the US Financial Markets

Giampiero M. Gallo and Edoardo Otranto

Abstract Empirical evidence shows that the dynamics of high frequency-based measures of volatility exhibit persistence and occasional abrupt changes in the average level. By looking at volatility measures for major indices, we notice similar patterns (including jumps at about the same time), with stronger similarities, the higher the degree of company capitalization represented in the indices. We adopt the recent Markov Switching asymmetric multiplicative error model to model the dynamics of the conditional expectation of realized volatility. This allows us to address the issues of a slow moving average level of volatility and of different dynamics across regimes. An extension sees a more flexible model combining the characteristics of Markov Switching and smooth transition dynamics.

1 Introduction

Direct measures of financial volatility were made possible by the availability of ultra high frequency data: several estimators were developed (for a review, see [1]) under a number of assumptions on the underlying continuous time process driving prices. In what follows, we will use the version called realized kernel volatility, proposed by [2], shown to filter out the presence of market microstructure noise and jumps.

G.M. Gallo
Department of Statistics "G. Parenti", Universitè di Firenze, Florence, Italy
e-mail: gallog@ds.unifi.it

E. Otranto (✉)
Department of Cognitive Sciences, Educational and Cultural Studies, Universitè di Messina, Messina, Italy
e-mail: eotranto@unime.it

M. Grigoletto et al. (eds.), *Complex Models and Computational Methods in Statistics*,
Contributions to Statistics, DOI 10.1007/978-88-470-2871-5_11,
© Springer-Verlag Italia 2013

When put next to one another, financial market volatilities generally exhibit similar behavior, being also subject to sudden, seemingly common, changes. Whether patterns of spillover can be detected from one market to another is the object of a large debate in the literature (see, for example, [6, 7], and references therein) which extends to the consequences of capital markets integration for portfolio diversification. In this paper we want to model volatilities in a univariate context with the aim to identify which indices present common features in the dynamics, given that each of them represents a different degree of market capitalization. The econometric approach is an extension of the multiplicative error model pioneered by [4] in the direction of identifying regimes of volatility with a Markov Switching behavior. Sudden changes typically occur when large shocks hit the markets and possibly showing up in the series as common to several indices. This impression is confirmed by the graphs in Fig. 1 where volatilities of six US indices are plotted in the period between January 3, 1996 and February 27, 2009: Standard & Poor's 500 (S&P500, 3263 obs.), Dow Jones 30 (DJ30, 3261 obs.), the S&P400 Midcap (S&P400, 3258 obs.), Russell1000 (RU1, 3262 obs.), Russell2000 (RU2, 3264 obs.) and Russell3000 (RU3, 3262 obs.).[1]

The visual inspection of these time series reveals a high degree of persistence and several abrupt changes, particularly clear in the most recent period, with turbulence leading to the burst of the tech bubble, the 2001 recession, the low level of volatility in the mid-decade and then the explosion of uncertainty following the subprime mortgage crisis. On the other hand, these peaks seem less marked, especially in the first part of the series, for S&P400 and RU2, which are indices representing companies with a lower degree of capitalization.

Recently, [8] has conjectured the presence of changing levels of the prevailing average volatility by subperiods: the series show in fact alternating regimes which visually involve changes in the level but may also correspond to differences in the dynamics in the series. They propose to extend the class of Multiplicative Error Models (MEMs), developed by [4] and expanded by [5], including a Markov Switching dynamics in the parameters to capture the presence of regimes. Being a MEM, this class of models applies to nonnegative-valued processes, therefore capturing dynamics without resorting to logs and producing forecasts of volatility (and not of log-volatility); moreover, considering the presence of regimes, these models capture the different phases of volatility, characterized by quiet periods, turmoil phases and accommodating brief abnormal peaks, leading to more realistic interpretations. In particular, applying their model to the same S&P500 volatility series analyzed here, [8] shows that it is possible to obtain a better fit relative to the standard MEM, to avoid the high persistence in the estimated series (which contrasts with the empirical evidence) and to eliminate the residual autocorrelation which affect many realized volatility models. In this paper we propose a further extension

---

[1]Data are expressed as percentage annualized volatility, i.e. the square root of the realized variance series taken from the *Oxford-Man Institute's realised library* version 0.1 [10], and multiplied by $\sqrt{252} * 100$.

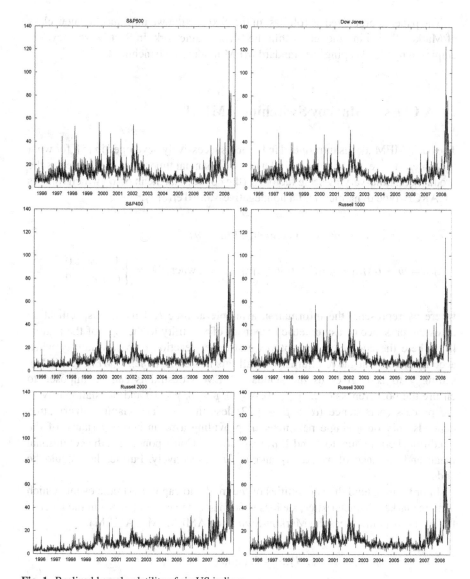

Fig. 1 Realized kernel volatility of six US indices

of this class of models, allowing also for the possibility that the parameters relative to the error distribution can follow a different change in regime than the parameters of the conditional expected volatility. This is achieved by considering, as in [11], smooth transition dynamics for the error coefficients, along the lines of [3]; for some series this extension improves the model performance. We select the best model in this class judging upon its statistical properties and drawing some considerations about the similarities in the changes in regimes across the six series.

The paper is organized as follows: in the next section we introduce the new class of Markov Switching models within the MEM framework. In Sect. 3 we show the empirical results, keeping the standard MEM model as a benchmark.

2   A Class of Markov Switching AMEM

The basic MEM idea is introduced in [4] and successively developed in [5]: for what is of interest here, the volatility x_t of a certain financial time series is modeled as the product of a time-varying scale factor μ_t (the conditional mean of x_t) which follows a GARCH-type dynamics, and a nonnegative-valued error ε_t:

$$x_t = \mu_t \varepsilon_t, \qquad \varepsilon_t | \Psi_{t-1} \sim \mathrm{Gamma}(a, 1/a) \quad \forall t$$

$$\mu_t = \omega + \alpha x_{t-1} + \beta \mu_{t-1} + \gamma D_{t-1} x_{t-1}, \qquad \text{where } D_t = \begin{cases} 1 \text{ if } r_t < 0 \\ 0 \text{ if } r_t \geq 0 \end{cases} \tag{1}$$

where Ψ_t represents the information available at time t. This *base* specification takes the presence of asymmetric responses of volatility to the sign of the returns [5], where the coefficient γ captures a stronger reaction to past volatility when accompanied by negative returns. We call this model *Asymmetric MEM* (AMEM); setting γ to zero gives us the standard MEM. Constraints can be imposed to ensure the positiveness of μ_t ($\omega > 0, \alpha \geq 0, \beta \geq 0, \gamma \geq 0$) and the stationarity of the process (persistence ($\alpha + \beta + \gamma/2$) less than 1). The Gamma distribution depends only on a single parameter a, providing a mean and a variance of the conditional error equal to 1 and $1/a$, respectively. Correspondingly, the conditional mean and variance of x_t are μ_t and μ_t^2/a, respectively. Further lags could be added.

In order to extend the capabilities of the model to capture extreme events which change market characteristics, such as sudden and persistent changes in the level of the series, [8] introduces the Markov-Switching AMEM (MS–AMEM):

$$x_t = \mu_{t,s_t} \varepsilon_t, \qquad \varepsilon_t | \Psi_{t-1} \sim \mathrm{Gamma}(a_{s_t}, 1/a_{s_t}) \quad \forall t$$

$$\mu_{t,s_t} = \omega + \sum_{i=1}^{n} k_i I_{s_t} + \alpha_{s_t} x_{t-1} + \beta_{s_t} \mu_{t-1,s_{t-1}} + \gamma_{s_t} D_{t-1} x_{t-1} \tag{2}$$

where s_t is a discrete latent variable which ranges in $[1, \ldots, n]$, representing the regime at time t. I_{s_t} is an indicator equal to 1 when $s_t \leq i$ and 0 otherwise; $k_i \geq 0$ and $k_1 = 0$. Accordingly, the constant in regime j is given by $(\omega + \sum_{i=1}^{j} k_i)$. The changes in regime are driven by a Markov chain, such that:

$$Pr(s_t = j | s_{t-1} = i, s_{t-2}, \ldots) = Pr(s_t = j | s_{t-1} = i) = p_{ij}. \tag{3}$$

Also in (2) the positiveness and stationary constraints given for (1) hold within each regime. Reference [8] identifies three regimes for the S&P500 realized kernel volatility that can be interpreted as the low, medium-high, and very high volatility states. Dealing with the same S&P500 series and in order to compare the changes in regimes of this series with other five series with similar dynamics, we also fix $n = 3$. When $\gamma_{s_t} = 0$, no asymmetric effects are present (MS–MEM).[2]

Accordingly, the unconditional mean of the volatility within each regime is:

$$m_{s_t} = \frac{\omega + \sum_{i=1}^{n} k_i I_{s_t}}{1 - \alpha_{s_t} - \beta_{s_t} - \gamma_{s_t}/2}. \tag{4}$$

The hypothesis that all the coefficients follow the same Markovian dynamics could be quite restrictive; for example, it would be plausible to think that the coefficient of the Gamma distribution follows its own dynamics not subject to the same regime changes as the coefficients of the conditional mean μ_t. We propose an alternative model to be used when the MS–AMEM does not fit the data adequately: we add another equation to (2) for the time-varying parameter of the Gamma distribution which changes more or less abruptly, depending on the value of the returns:

$$a_t = b_0 + b_1 \{1 + \exp[-\delta(r_{t-1} - c)]\}^{-1} \tag{5}$$

where $b_0 > 0$, $b_1 \geq 0$, $\delta > 0$ and c are unknown parameters. In practice, we are adding a time-varying smooth transition variance (see [13]), not dependent on regimes, but with a suitable dynamic behavior. We call the model (2)–(3)–(5), the MS–AMEM with Smooth Transition Variance (MS–AMEM–STV). This specification would provide more flexibility to the Markov Switching model, in particular to capture the sizeable jumps, such as the highest peaks in 2008 (see Fig. 1). When δ approaches ∞, (5) is equivalent to a threshold model [14], and (5) is substituted by:

$$a_t = \begin{cases} b_0 & \text{if } r_{t-1} \leq c \\ b_0 + b_1 & \text{if } r_{t-1} > 0 \end{cases}. \tag{6}$$

In this case we obtain different regimes for the conditional mean equation and for the Gamma coefficient, which will follow proper dynamics with two regimes. We call the model represented by (2)–(3)–(6), the MS–AMEM with Threshold Variance (MS–AMEM–TV).

---

[2]Details about the reparameterization of β_{s_t} to guarantee a certain coherence between the regime and the level of volatility, and about the solution of possible estimation problems, are in [8]. In the same work another specification of the MS–AMEM is given, in which the asymmetry deriving from the sign of the returns may affect also the transition probabilities (the so-called *Asymmetry in Probability MS–AMEM*).

3 Empirical Results

As a complement to the profile of the six volatility series shown in Fig. 1 above, we have calculated the usual descriptive statistics (not shown here to save space). They confirm the compatibility with the presence of regimes, especially a very large range with a thick right tail (high kurtosis). Time dependence is reflected in the autocorrelation functions, which are characterized by slowly declining high values, a fact typically seen as evidence of the presence of regimes.

We estimate the MS–AMEMs for all the series and verify if they have a good performance in terms of fitting and statistical tests; in particular we adopt the autocorrelation pattern of the residuals as a guideline, in the sense that, if they are correlated, we estimate the alternative MS–AMEM–STV (the MS–AMEM–TV if δ diverges) choosing the one with better properties in terms of results of the Ljung-Box statistics. This procedure selects the MS-AMEM only for the S&P500 volatility (as shown in detail in [8]), the MS–MEM for the DJ30 series, the MS–AMEM–STV for the S&P400 volatility, and the MS–AMEM–TV for the three Russell indices.

We have also estimated the original AMEM, shown in (1), for all the series and compared its statistical performance with respect to the selected MS models. For this purpose we calculated the AIC[3] and some loss functions of interest, namely, the Root Mean Squared Error (RMSE), the Mean Absolute Error (MAE) and Theil's U (the latter calculated using the first differences of observed and forecasted data to detect the capability of the model to capture the turning points). On all accounts Markov Switching behavior is detected (see Table 1), with a strong improvement in the residual diagnostics.

One of the motivations to adopt an MS volatility model is the presence of autocorrelated residuals in the AMEM. In the same Table 1 we show the p-values of the Ljung-Box test statistics, in correspondence of lags 1, 5, and 10, for the AMEM and the selected MS models to check how uncorrelated the residuals are.[4] What we observe is that the models with three regimes are able to capture a large portion of the strong residual dependence structure still present in the AMEM.

The estimation results for the MS models are reported in Table 2. We notice a strong difference in model dynamics when the assumption of common dependence of the coefficients on the regimes is relaxed. Starting from the intercepts, the MS models show a significant increase in these coefficients when regimes change, with

---

[3]Tests based on the likelihood function cannot be used to compare the AMEM with respect to the corresponding MS models because of the presence of nuisance parameters present only under the alternative hypothesis; in this case, with the proper caution, a classical information criterion could provide some information (see [12]); in particular the AIC seems to choose the correct state dimension more successfully than the BIC, provided that the parameter changes are not too small and the hidden Markov chain is fairly persistent.

[4]For MS models we have used the generalized residuals, introduced by [9] for latent variable models, defined as $E(\hat{\varepsilon}_t|\Psi_{t-1}) = \sum_{i=1}^{3} \hat{\varepsilon}_{s_t,t} tPr(s_t = i|\Psi_{t-1})$, where $\hat{\varepsilon}_{s_t,t}$ are the residuals at time t derived from the parameters of the model in state s_t.

Table 1 Likelihood-based criteria, in-sample forecasting performance and autocorrelation tests[a] for AMEM and MS models

	Log-lik	AIC	RMSE	MAE	Theil U	$p(Q_1)$	$p(Q_5)$	$p(Q_{10})$
S&P500								
AMEM	−8389.66	5.145	4.490	2.811	0.381	0.002	0.000	0.002
MS–AMEM	−8328.77	5.118	4.428	2.632	0.367	0.140	0.027	0.103
DJ30								
AMEM	−8056.77	4.944	4.059	2.478	0.370	0.002	0.000	0.000
MS–MEM	−8025.82	4.933	4.074	2.330	0.361	0.809	0.008	0.091
S&P400								
AMEM	−7516.42	4.617	3.528	2.197	0.339	0.001	0.000	0.005
MS–AMEM–STV	−7467.62	4.598	3.312	1.929	0.295	0.140	0.014	0.011
RU1								
AMEM	−8158.18	5.005	4.194	2.620	0.378	0.002	0.000	0.002
MS–AMEM–TV	−8100.88	4.980	3.716	2.303	0.334	0.270	0.017	0.023
RU2								
AMEM	−7462.25	4.576	3.717	2.245	0.376	0.003	0.000	0.003
MS–AMEM–TV	−7423.91	4.562	3.515	1.970	0.325	0.169	0.005	0.001
RU3								
AMEM	−8054.88	4.942	4.081	2.542	0.376	0.001	0.000	0.001
MS–AMEM–TV	−8013.71	4.927	3.668	2.260	0.335	0.146	0.036	0.052

Sample: January 3, 1996 to February 27, 2009
[a]In the table, $p(Q_j)$ $(j = 1, 5, 10)$ indicates the p-values of the Ljung-Box test statistics at lag j

an increase of more than 5 points in the high volatility third regime; the models with MS and STV or TV do not show similar differences in the intercepts. In this case the more flexible variance is able to capture also abrupt jumps in the series maintaining small intercepts. As a consequence, volatility dynamics is represented by different coefficient behavior; in the case of models type (2), the α and γ coefficients increase with the regime whereas the β coefficients show an opposite behavior; this involves a strong dependence on the most recent observation and on the sign of returns for the regimes of high volatility and a lower persistence. The models containing (6) show that the third regime depends only on the values corresponding to negative returns and an increasing persistence in the third regime. It is interesting to note also that the estimated coefficients of the RU1 and RU3 volatility are very similar (pointing to a common DGP), whereas they differ from the one of RU2. We can argue that the companies with larger capitalization present in both RU1 and RU3 dominate the behavior of the volatility, while the smaller caps in RU2 behave differently.

In terms of transition probabilities, it is evident that there is a strong permanence in the same regime for all the indices, in particular in regime 1 and 2. Regime 3 is less persistent for all Russell's and for S&P400. Some further insights are gained by looking at the off-diagonal elements of the transition probability matrix, with similar considerations for all the indices. Being in regime 1 there is a very low probability to switch to either of the other two regimes. From the regime of intermediate volatility there is a higher probability to move to the high volatility regime than to revert to

Table 2 Coefficient estimates for Markov Switching AMEM specifications with three regimes (standard errors in parentheses). Sample: January 3, 1996 to February 27, 2009[a]

	S&P500	DJ30		S&P400	RU1	RU2	RU3
ω	1.872	2.367		0.911	1.652	0.843	1.637
	(0.229)	(0.927)		(0.018)	(0.166)	(0.121)	(0.151)
k_2	0.685	1.048		0.000	0.001	0.000	0.000
	(0.228)	(0.242)		(0.001)	(0.006)	(0.005)	(0.001)
k_3	5.188	5.990		0.989	0.827	0.901	0.717
	(1.326)	(2.775)		(0.002)	(0.117)	(0.082)	(0.334)
α_1	0.199	0.311		0.160	0.180	0.206	0.182
	(0.028)	(0.112)		(0.007)	(0.022)	(0.021)	(0.021)
α_2	0.161	0.270		0.073	0.098	0.055	0.101
	(0.029)	(0.096)		(0.003)	(0.011)	(0.010)	(0.012)
α_3	0.257	0.385		0.000	0.000	0.000	0.000
	(0.058)	(0.086)		(0.000)	(0.000)	(0.000)	(0.000)
β_1	0.525	0.377		0.637	0.563	0.554	0.559
	(0.056)	(0.224)		(0.008)	(0.043)	(0.051)	(0.041)
β_2	0.594	0.460		0.789	0.714	0.807	0.707
	(0.058)	(0.191)		(0.004)	(0.024)	(0.020)	(0.023)
β_3	0.343	0.230		0.932	0.896	0.928	0.876
	(0.108)	(0.171)		(0.009)	(0.063)	(0.013)	(0.031)
γ_1	0.076			0.042	0.077	0.035	0.077
	(0.007)			(0.010)	(0.010)	(0.023)	(0.010)
γ_2	0.083			0.067	0.091	0.083	0.091
	(0.010)			(0.014)	(0.008)	(0.006)	(0.009)
γ_3	0.143			0.068	0.104	0.072	0.124
	(0.018)			(0.009)	(0.062)	(0.013)	(0.031)
a_1	15.808	21.062	b_0	4.314	5.226	3.945	5.012
	(0.632)	(1.249)		(1.554)	(1.947)	(1.259)	(2.247)
a_2	18.742	20.677	b_1	14.537	11.249	11.568	11.256
	(1.452)	(1.675)		(1.557)	(2.068)	(1.518)	(2.280)
a_3	10.946	11.127	δ	0.936			
	(0.751)	(1.194)		(0.021)			
			c	−3.531	−3.636	−4.310	−3.630
				(0.734)	(0.022)	(0.021)	(0.008)
p_{11}	0.989	0.977		0.994	0.992	0.985	0.993
	(0.002)	(0.009)		(0.002)	(0.001)	(0.004)	(0.001)
p_{12}	0.007	0.018		0.000	0.004	0.010	0.004
	(0.001)	(0.004)		(0.000)	(0.001)	(0.006)	(0.001)
p_{13}	0.004	0.005		0.006	0.004	0.005	0.003
p_{21}	0.007	0.013		0.000	0.003	0.003	0.004
	(0.001)	(0.005)		(0.000)	(0.002)	(0.002)	(0.001)
p_{22}	0.977	0.975		0.951	0.945	0.948	0.950
	(0.002)	(0.003)		(0.002)	(0.016)	(0.009)	(0.009)
p_{23}	0.016	0.012		0.049	0.052	0.049	0.046
p_{31}	0.006	0.007		0.007	0.014	0.013	0.011
	(0.002)	(0.012)		(0.001)	(0.009)	(0.008)	(0.004)
p_{32}	0.042	0.049		0.247	0.210	0.203	0.170
	(0.001)	(0.008)		(0.013)	(0.097)	(0.019)	(0.042)
p_{33}	0.952	0.944		0.746	0.776	0.784	0.819

[a]The model selected are: an MS–AMEM for S&P500, an MS–MEM for DJ30, an MS–AMEM–STV for S&P400, an MS–AMEM–TV for RU1, RU2 and RU3. The coefficients p_{i3} ($i = 1, 2, 3$) are not directly estimated, but are obtained as $p_{i3} = 1 - p_{i1} - p_{i2}$

Table 3 Unconditional mean of the volatility within each regime of US financial indices in the period from January 3, 1996 to February 27, 2009

	S&P500	DJ30	S&P400	RU1	RU2	RU3
m_1	7.891	7.599	5.011	7.561	3.780	7.424
m_2	12.609	12.648	8.551	11.557	8.719	11.184
m_3	23.570	24.442	55.937	47.675	48.513	37.986

a low volatility regime. By the same token, we note that the downward transition from the high volatility states occurs preferably with a move to the intermediate state: joint with the considerations above, there seems to be a strong interaction between regimes 2 and 3 while the period of low volatility is a sort of self-standing regime.

The different behavior of the previous coefficients could be misleading in the interpretation of regimes; the level of the volatility within each regime is represented by the unconditional mean (4) by regime, which is shown in Table 3 for each series and signals the interpretability of regimes as increasing volatility. Moreover, S&P500, DJ30, RU1, and RU3 present similar levels of volatility in regime 1 and 2, which are higher with respect to the corresponding levels of S&P400 and RU2. The third regime is the one presenting the main differences among the six series; S&P500 and DJ30 are again similar, whereas S&P400, RU1 and RU2 show very high levels of average volatility, with RU3 in an intermediate position. Using smoothed probabilities, we can superimpose the average volatility levels to the observed series as in Fig. 2. Bursts of volatilities, as well as sudden reductions in their values, correspond to a discrete change in the average value around which volatility follows its dynamics. More erratic behavior is apparent in the less frequently inspected indices. It is clear that the MS–AMEM with STV or TV consider the third regime as a state which absorbs the highest peaks, whereas the MS–AMEM corresponds to a higher duration in the regime.

In practice, it seems that there is a certain consistency in the behavior of the high capitalization indices, whereas S&P400 and RU2 show a sort of definitive permanent level shift at the end of 1998, with consistently higher levels of volatility from there on. The coherence among the indices can be evaluated in Table 4, where we show the percentage of cases in which the indices fall in the same regime. The high capitalization indices are in the same regime in more than 82% of cases, with a maximum in correspondence of RU1 and RU3 (96%). The coherence between high and low capitalization indices is low (between 44% and 65%), whereas there is a high coherence between S&P400 and RU2 (88.5%).

Figure 3 addresses the issue of whether regimes are coherent across indices or whether the non-homogeneity is relative to a specific state: we build a bar graph where the frequency in each regime for one index is broken down by the frequency across its own regimes for another index. The S&P500 is the reference index in five panels while the last one reports results between S&P400 and RU2. If regimes

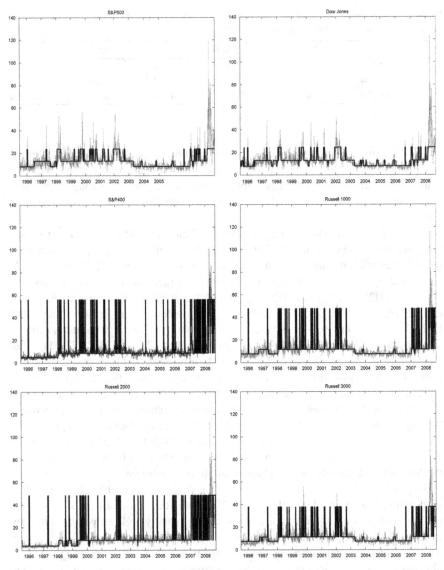

Fig. 2 Original series (*dotted lines*) with regime-specific average volatilities (*bold lines*) in the sample period January 3, 1996 to February 27, 2009

agreed perfectly, we would have each side bar of the color of the same regime. Variety signals different regime partition. The most striking result is the large coherence between S&P500 and DJ30 at one hand (top left panel) and between S&P400 and RU2 on the other (bottom right panel). The most striking contrast is between the latter two each with the S&P500 (right column, top and middle). This suggests that the small cap companies have a more similar behavior as the mid

		DJ30	S&P400	RU1	RU2	RU3
Table 4 Percentage of common regimes between pairs of realized volatilities in the period from January 3, 1996 to February 27, 2009	S&P500	89.26	52.71	86.45	44.47	85.74
	DJ30		52.30	83.15	45.23	82.20
	S&P400			62.54	88.49	64.68
	RU1				54.23	95.92
	RU2					56.74

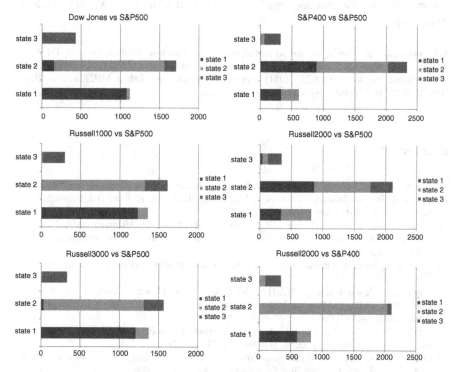

Fig. 3 Distribution of the regimes between pairs of financial US indices in the sample period January 3, 1996 to February 27, 2009. A vs B indicates that the frequency in each regime for index A is broken down by the frequency across its own regimes for index B

caps, while the largest of the large caps (well represented by the S&P500) dominate volatility behavior when inserted within an index.

4 Conclusions

With direct volatility measurement, many interesting questions can be addressed about its dynamics. We have investigated the possibility that abrupt changes seen in the time series of realized kernel volatility may signal the presence of regimes corresponding to different average levels of turbulence. With our Markov Switching specification of a multiplicative error model we have allowed for the possibility that the shape parameter of the Gamma distribution ruling the tails of the error term may be made dependent on the value of the lagged returns. This significantly adds to the catalog of available volatility models for forecasting. While run on individual series, the analysis allows to compare results and establish commonalities. Company size shows up in different behavior in the volatility of indices for large caps on the one side and for mid and small caps on the other, an issue which is seldom investigated.

Acknowledgments Thanks are due to participants in the following conferences: ERCIM 2010 (London, December 10–12, 2010), ECTS 2011 (Monte Porzio Catone, June 13–14, 2011), SCO 2011 (Padova, September 19–21, 2011). Financial support from Italian MIUR under Grant 20087Z4BMK_002 is gratefully acknowledged.

References

1. Andersen, T.G., Bollerslev, T., Diebold, F.X.: Parametric and nonparametric volatility measurement. In: Aït-Sahalia, Y., Hansen, L.P. (eds.) Handbook of Financial Econometrics, pp. 67–138. North-Holland, Amsterdam (2010)
2. Barndorff-Nielsen, O.E., Hansen, P.R., Lunde, A., Shephard, N.: Designing realised kernels to measure the ex-post variation of equity prices in the presence of noise. Econometrica **76**, 1481–1536 (2008)
3. Chan, K.S., Tong, H.: On estimating thresholds in autoregressive models. J. Time Ser. Anal. **7**, 178–190 (1986)
4. Engle, R.F.: New frontiers for ARCH model. J. Appl. Econ. **17**, 425–446 (2002)
5. Engle, R.F., Gallo, G.M.: A multiple indicators model for volatility using intra-daily data. J. Econ. **131**, 3–27 (2006)
6. Engle, R.F., Gallo, G.M., Velucchi, M.: Volatility spillovers in East Asian financial markets: a MEM-based approach. Rev. Econ. Stat. **94**, 222–233 (2012)
7. Gallo, G.M., Otranto, E.: Volatility spillovers, interdependence and comovements: a Markov switching approach. Comput. Stat. Data Anal. **52**, 3011–3026 (2008)
8. Gallo, G.M., Otranto, E.: The Markov switching asymmetric multiplicative error model. WP CRENoS **2012/05** (2012)
9. Gourieroux, C., Monfort, A., Trognon, E.R.A.: Generalized residuals. J. Econ. **34**, 5–32 (1987)
10. Heber, G., Lunde, A., Shephard, N., Sheppard, K.: OMI's Realised Library, Version 0.1. Oxford-Man Institute. University of Oxford, Oxford (2009) Available at http://realized.oxford-man.ox.ac.uk/data
11. Otranto, E.: Classification of volatility in presence of changes in model parameters. WP CRENoS **2011/13** (2011)
12. Psaradakis, Z., Spagnolo, F.: On the determination of the number of regimes in Markov-switching autoregressive models. J. Time Ser. Anal. **24**, 237–252 (2003)
13. Teräsvirta, T.: Recent developments in GARCH modeling. In: Andersen, T.G., Davis, R.A., Kreiß, J.-P., Mikosch, T. (eds.) Handbook of Financial Time Series, pp. 17–42. Springer, Berlin (2009)
14. Tong, H.: Non-Linear Time Series. Clarendon, Oxford (1990)

Semicontinuous Regression Models with Skew Distributions

Anna Gottard, Elena Stanghellini, and Rosa Capobianco

Abstract In applied studies, researchers are often confronted with semicontinuous response models. These are models with a semicontinuous response variable, i.e. a continuous variable that has a lower bound, that we here consider to be zero, and such that a sizable fraction of the observations takes value on this boundary. Semicontinuous response models are common in pharmacovigilance, pharmacoepi-demiological and pharmacoeconomic studies, where it can be sometimes useful to evaluate and monitor the doses of a certain drug substance consumed by the general population. The preponderant number of observations taking value zero corresponds to the part of the population which is not actually consuming the medicine, either because they do not need it or because, even if they need it, they are not taking it in a given interval of time. Another interesting field of application concerns goods or drugs consumption to be studied for economic or social purposes. We here explore the use of several asymmetric distributions to address the fact that the continuous part of the data distribution shows skewness in most cases. As an illustration, the proposal is applied to model alcohol expenditure in Italian households.

A. Gottard
Department of Statistics "G. Parenti", University of Florence, Florence, Italy
e-mail: gottard@ds.unifi.it

E. Stanghellini (✉)
Department of Economics, Finance and Statistics, University of Perugia, Perugia, Italy
e-mail: elena.stanghellini@stat.unipg.it

R. Capobianco
Studies on Intercultural, Cultural and Training Processes in Contemporary Society, Roma Tre University, Rome, Italy
e-mail: rcapobianco@uniroma3.it

M. Grigoletto et al. (eds.), *Complex Models and Computational Methods in Statistics*, 149
Contributions to Statistics, DOI 10.1007/978-88-470-2871-5_12,
© Springer-Verlag Italia 2013

1   Introduction

In many situations it can be of interest to model a response variable having an unexpected, with respect to a given distribution, number of zero values. For count responses, a situation of excess of zeroes with respect to a standard distributional assumption, such as Poisson or Negative Binomial distribution, is usually handled in the literature by adopting a so-called zero-inflated model [13] or a hurdle model [16]. An ordinary Poisson or Negative Binomial regression model would be inadequate to study these kinds of data. See [14] for a survey on these classes of models.

Continuous positive value data with a preponderance of zero observations can occur in a variety of applications. These kinds of variables are sometimes called semicontinuous [17] to allow a probability mass on a specific value. Semicontinuous data are common in many fields. For example, in pharmacovigilance, pharmacoepidemiological and pharmacoeconomic studies, it can be sometimes useful to evaluate and monitor the doses of a certain drug consumed in the general population, outside of a clinical trial. Also, in many therapeutic fields, as nonsteroidal anti-inflammatory drugs, oral contraceptives, antihypertensive drugs, antidepressants, and so on, the actual drug consumption is an important item of the total public expenditure. The preponderant presence of zeroes observed in these frameworks corresponds to the part of the population which is not actually consuming the medicine, either because they do not need it or because they are not taking it in the interval time of monitoring, even if needed. Similarly, in meteorological studies, the daily quantity of rainfall or snowfall distribution can present a clump at zero, due to not rainy/snowy days. However, due to some measurement errors, small quantity of rain in rainy days may not be detected. A further, interesting field of application is measuring goods, food or drug consumption or expenditure. In this paper, the case of alcohol expenditure in Italian households is considered. The distribution of alcohol consumption presents a clump at zero, due to both teetotalers and occasional drinkers having no alcohol during the monitoring period.

The first model for semicontinuous data is the so-called *Tobit model*, due to [21]. The Tobit model assumes an underlying normally distributed latent variable determining when the outcome of interest is positive or zero. Since the assumption of normality of the underlying latent variable is too restrictive in many applications, several extensions of the Tobit model have been proposed in the literature, especially in the direction of distribution-free methods. See [18] for a summary. Also, as noted in several applied studies, the continuous part of the data distribution may present a rather substantial skewness. To address this issue, [11] assumes a Skew-Normal model [3] for the latent variable that governs the outcome of interest, for the case of clumping due to right censoring in bounded health scores.

The two-part model [8, 9] extends the Tobit model to the case when the probability of zero and positive values depends on different sets of parameters and explanatory variables. The likelihood function of this class of models factorizes into two parts that can be maximized separately. According to [14], the two-part model is sometimes preferable as it addresses the data in their original form, it is simple to fit and to interpret. It can be seen as the continuous counterpart of the hurdle model.

A different generalization of the Tobit model, used in microeconometric studies, is called *Double-Hurdle* model (see, for example, [4, 8, 12]). This kind of model has been used to study individual behavior on commodity consumptions as two subsequent decisions. Therefore, two separate hurdles have to be cleared in order to obtain a positive level of consumption. The Double-Hurdle model allows a dependence between the two components.

The two-part model has been extended in several ways (see, for example, [6, 15, 20]). As for the Tobit model, the continuous part of the data can be skewed, as shown in several studies. This issue is generally addressed by taking the log of the response variable. However, this can lead to a normalization of only lightly skewed situations. In particular, [6] proposes a probit/log-skew-normal distribution to analyze continuous data with a discrete component at zero.

In this paper, we are considering a two-part model combining several models for both the discrete component and the continuous component with skew distributions. This kind of model is able to account for several forms of asymmetry. In particular, the Skew-Normal and the Skew t distributions for the continuous part can accommodate asymmetry in a very flexible way. The first model reduces to the normal model when a particular parameter, denoted by α, equals zero. For this case, the score functions have been derived and it is shown that, when $\alpha = 0$, the score functions are linearly dependent so that the information matrix is singular.

The paper is organized as follows. The proposed two-part model is presented in Sect. 2. Some properties of the proposed class of models are given in Sect. 3. The analysis of alcohol consumption data is in Sect. 4. Section 5 concludes with some remarks. The information matrix of a particular class of models here proposed is derived in the Appendix.

2 Semicontinuous Response Models Using Skew Distributions

Let Y_1 and Y_2 be two independent latent random variables. Assume Y_1 follows a Bernoulli distribution, with

$$P(Y_1 = 1) = \pi_1 \quad \text{and} \quad P(Y_1 = 0) = \pi_0 = 1 - \pi_1.$$

Let Y_2^* be a further latent random variable with support on $\mathbb{R}$. Define Y_2 to have a left censored at zero distribution with support on $[0, +\infty)$. We suppose that the probability distribution of Y_2 is linked to that of Y_2^* as

$$P(Y_2 = 0) = P(Y_2^* \leq 0) \quad \text{while} \quad P(Y_2) = P(Y_2^*) \quad \text{when} \quad Y_2^* > 0.$$

Finally, define Y as the observable random variable with the clump at zero and a continuous positive part. We assume:

$$Y = \begin{cases} 0 & \text{if} \quad Y_1 = 0 \\ Y_2 & \text{if} \quad Y_1 = 1 \quad \text{with } Y_2 \geq 0. \end{cases} \tag{1}$$

Therefore,

$$P(Y = 0) = P(Y_1 = 0) + P(Y_1 = 1, Y_2 = 0) = P(Y_1 = 0) + P(Y_1 = 1)P(Y_2^* \leq 0)$$

as Y_1 and Y_2 are assumed independent and $P(Y_2 = 0) = P(Y_2^* \leq 0)$. It follows that the density function of the random variable Y, say f_Y, can be written as

$$f_Y(y) = [\pi_0 + \pi_1 P(Y_2^* \leq 0)] I(y = 0) + \pi_1 f_{Y_2}(y) I(y > 0). \tag{2}$$

Note that two different mechanisms may give rise to $Y = 0$. One is driven by the latent variable Y_1, while the other is driven by the truncation of the latent variable Y_2. In the pharmacoeconomic studies the Y_1 variable describes the need or not to take a given drug, while Y_2 drives the actual consumption for those in need to take the drug. In an actual study, Y_2^* may be zero, especially if the monitoring time is short.

Moreover, notice that, if $\pi_0 = 0$, the model is a left-censored-at-zero model. Whenever $\pi_0 > 0$, the proposed model assumes an extra source of zero values due to the discrete component Y_1. Furthermore, if $P(Y_2^* \leq 0) = 0$, and consequently Y_2 has a strictly positive support, then the resulting model is a two-part model. Otherwise, the model can be viewed as a zero-inflated truncated continuous model. See for alternative proposals [1, 7]. For Gaussian specification of the continuous part of the model, our proposal coincides with the Double-Hurdle model with independent components.

As already mentioned, we may want the distribution of the continuous part of the data to allow for skewness. Several possible choices are available. Here we shall consider two possible choices: (a) the Skew-Normal distribution and (b) the Skew t distribution.

3 The Specification of the Model

As illustrated by Lemma 1 of [2], a skew distribution may be generated by perturbation of a symmetric distribution, as follows:

$$g(z) = 2f_0(z)G_0(wz) \tag{3}$$

where f_0 is a one-dimensional density function, being symmetric about 0, and G_0 is a one-dimensional distribution function such that G_0 exists and has a density symmetric about 0, while $w(\cdot)$ is an odd function.

Under choice (a), when $f_0 = \phi(z)$ and $G_0 = \Phi(\alpha z)$, where $\phi(\cdot)$ and $\Phi(\cdot)$ are, respectively, the density and the distribution function of a standard univariate normal distribution, (3) is the density function of a Skew-Normal random variable, with skewness parameter α, $Z \sim SN(\alpha)$.

In the model in (2) under choice (a), we shall impose $Y_2^* = \mu + \sigma Z$. In this case, $Y_2^* \sim SN(\mu, \sigma^2, \alpha)$. Notice that for $\alpha = 0$ the model reduces to $Y_2^* \sim N(\mu, \sigma^2)$. The effect of covariates on the parameter μ can also be investigated. A possible choice is to assume for the i-th generic observation:

$$\mu_i = \beta_2' X_{2i} \tag{4}$$

where X_{2i} denotes a set of explanatory variables for Y_2^* and β_2 is a vector of regression coefficients.

Choice (a) parallels what done by [6] in which a log transformation is assumed instead, so that the continuous part has strictly positive support. As shown in the Appendix, a test for $\alpha = 0$, which corresponds to testing for the absence of skewness in the data, may be problematic. This is also true for right censored data (see [5]).

Under choice (b) of a continuous component with Skew t distribution, we let $f_0 = t(z; \nu)$ and

$$G_0 = T\left[\alpha \sigma^{-1}(y - \mu)\left(\frac{\nu + 1}{Q + \nu}\right)^{1/2}; \nu + 1\right]$$

where $Q = [(y - \mu)/\sigma]^2$, t and T are, respectively, the density and the distribution function of a Student t distribution with ν and $\nu + 1$ degrees of freedom. In a similar manner as in (4), we shall impose that $Y_2^* = \mu + \sigma Z$ to account for the effect of covariates on μ.

Focusing now on assumptions on Y_1, we may consider the effect of covariates also on π_1, by the insertion of a suitable link function. Classical choices of link function include logit or Probit model.

To exemplify, under choice (a) of a Skew-Normal distribution for Y_2^*, and a logit link for Y_1, the contribution to the likelihood function of a generic observation i is

$$L_i^{L.SN} = \left[\frac{1}{1 + \exp \beta_1' x_{1i}} + \frac{\exp \beta_1' x_{1i}}{1 + \exp \beta_1' x_{1i}} F_{SN}\left(0; \beta_2' x_{2i}, \sigma^2, \alpha\right)\right]^{I[y_i = 0]} \cdot$$

$$\left[f_{SN}\left(y_i; \beta_2' x_{2i}, \sigma^2, \alpha\right)\right]^{I[y_i = 1]}$$

where $f_{SN}(\cdot)$ and $F_{SN}(\cdot)$ are the density and the distribution function of the Skew-Normal variable. Assuming a probit link, under choice (a), the contribution to the likelihood function of a generic observation i is instead

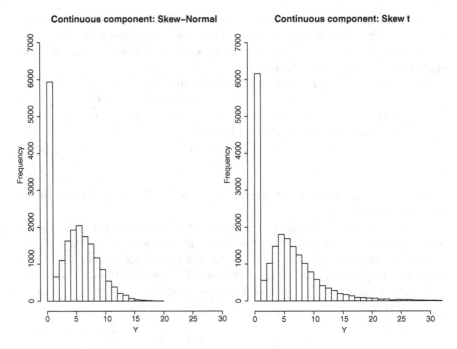

Fig. 1 Simulated data for variable Y, under choice (a) and (b) with a Probit link

$$L_i^{P.SN} = \left[\left(1 - \Phi(\beta_1' x_{1i})\right) + \Phi(\beta_1' x_{1i}) F_{SN}\left(0; \beta_2' x_{2i}, \sigma^2, \alpha\right) \right]^{I[y_i=0]} \cdot$$

$$\left[f_{SN}\left(y_i; \beta_2' x_{2i}, \sigma^2, \alpha\right) \right]^{I[y_i=1]} \cdot$$

As an example, we report in Fig. 1 the histogram of two data sets randomly generated by model (2) under choice (a) and (b) with the probit link. The data are generated assuming for both the distributions $\pi_1 = 0.6$ and the location, scale and skewness parameters equal to 6, 3 and 0.5, respectively. For the Skew t distribution the number of degrees of freedom was settled at 3. It can be seen that the Skew t choice yields to a heavier right tail, giving the possibility of adequately model the presence of kurtosis in the data. Notice that, even if the parameter π_1 is the same, the amount of zero values in the Skew t case is lightly higher, due to truncation of the left tail, heavier than that of the Skew-Normal distribution. However, the distribution under choice (b) tends to that under choice (a) as the number of degrees of freedom tends to infinity.

4   Analysis of the Alcohol Consumption Data

As an illustration, we analyze here a set of data on alcohol consumption coming from the 2002 Italian Household Budget Survey (IHBS) conducted by Istat. The sample consists of 27,499 Italian households. Expenditure on a wide range on nondurable goods and services over 1 week period is recorded and subsequently expressed on a monthly basis. Expenditure variables are deflated via regional price indices produced by Istat. Data refer to the households rather than individuals, therefore a per-equivalent adult measure is obtained by adjusting for family size using the modified OECD equivalence scale, which assigns weight 1 to the first adult, 0.5 to each subsequent adult, and 0.3 to each child under 14 years of age.

Here the interest is in the determinants of monthly expenditure for alcohol. Explanatory variables to take into account for socioeconomic factors of the household are recorded. Also the amount of expenditure in tobacco is recorded, a variable that emerges as significant in our analysis. In order to amplify the differences between the continuous and the discrete part of the distribution, the response variable has been transformed using $Y = \log(1 + X)$. This is usually done in studies on semicontinuous response models (see, e.g., [6]). The histogram of the so transformed response variable, in Fig. 2, shows that there is a moderate evidence of skewness in the positive data.

The probit link has been chosen for the discrete component. For the continuous part, we adopted the Skew-Normal distribution (choice (a) in the previous sections). The model has been selected starting from a large model. Included covariates are: age (age), with a linear and quadratic effect, percentage of adult male members of the household (perc_males), income proxied by per-equivalent adult household total expenditure scaled by 100 (income), per-equivalent tobacco consumption categorized in tertiles (ftab), a binary variable taking value 1 if there is at least one child aged 0–14 years (Child014) and list of binary variables taking value 1 if the household's head: is male (MaleH), has a higher education (HighEdu), has a white collar job position (Whitecollar), owns the house (OwnerOcc), is single without children (Single).

The first model included all variables both in the Probit link and in the continuous part of the model (with the exception that the variable income has been categorized in four levels (fincome), using quartiles as cut-points, for the continuous component to avoid instability of estimates and allow for nonlinear effects). Using a backward selection procedure, we proceeded by removing variables with nonsignificant effect. The estimates of the final model are in Table 1. The likelihood ratio test between the extended and the reduced model was 11.13 with 10 degrees of freedom.

Variables have an effect in the expected direction. Tobacco consumption plays a role both in the decision model (Probit equation) and in the level model (Skew-Normal equation). If household's head has a white collar job the probability of drinking alcohol (binary component) reduces, while it increases if the household's head is male. These binary covariates, however, do not affect the continuous component. The main negative effect on the level of alcohol expenditure is the

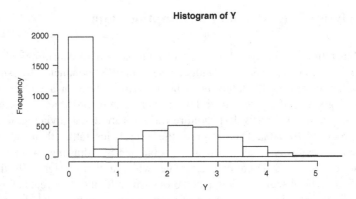

Fig. 2 Observed values for variable Y, monthly expenditure for alcohol

Table 1 Parameter estimates for the selected model

Variables	Estimate	s.e.	Confidence intervals (95%)	
Model for the binary component				
Intercept	−1.3686	0.1118	−1.5878	−1.1494
eta	0.0168	0.0021	0.0128	0.0209
eta2	−0.0005	0.0002	−0.0008	−0.0002
perc_admales	0.7036	0.1017	0.5042	0.9029
income	0.0104	0.0019	0.0066	0.0141
MaleHH	0.3176	0.0658	0.1887	0.4466
whitecollar	−0.1323	0.0409	−0.2124	−0.0523
ftab(0.1,20]	0.4700	0.0530	0.3661	0.5739
ftab(20,160]	0.3610	0.0516	0.2599	0.4622
Model for the continuous component				
Intercept	1.8089	0.0582	1.6949	1.9229
perc_admales	0.2618	0.0762	0.1125	0.4112
fincome(6.2,9.08]	0.3198	0.0500	0.2217	0.4179
fincome(9.08,13.3]	0.4741	0.0499	0.3762	0.5720
fincome(13.3,167]	0.6797	0.0509	0.5799	0.7794
single	0.3400	0.0539	0.2343	0.4458
child014	−0.1197	0.0360	−0.1903	−0.0491
ftab(20,160]	0.1466	0.0410	0.0663	0.2270
s.d.	0.8120	0.0122	0.7881	0.8358
skewness	0.1766	0.0587	0.0614	0.2917

presence in the family of children aged less than fourteen. Finally, income positively affects alcohol expenditure in both components.

Notice that the derivation in the Appendix implies that asymptotic distribution of the likelihood ratio test to compare normal and skew Normal models is nonstandard. However, the skewness parameter is significant, even if small. Parameter estimates and Hessian matrix have been obtained by a numerical algorithm in the optim function implemented in R [19]. Attempts to use the Skew t distribution have been

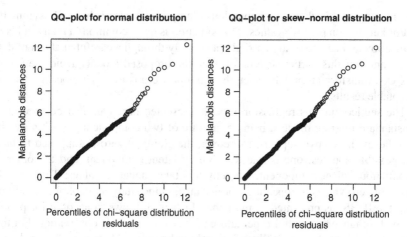

Fig. 3 Healy's plot of the residuals of the nonzero values correctly identified against (*left panel*) the normal distribution and (*right panel*) the Skew-Normal distribution

made; however, the algorithms for the maximization of the log likelihood did not converge. As far as it concerns the algorithm for likelihood maximization in the Skew-Normal case, the more stable results have been obtained using simulated annealing. This algorithm is a global optimization method. It is quite inefficient compared to the classical Newton–Raphson-like methods but it is able to distinguish among several local optima. This characteristic makes it quite convenient in case of difficult log-likelihood functions.

To evaluate the model fitting, it is interesting to first compare fitted and observed values in terms of the binary component Y_1. For this purpose, for each unit i we set $\hat{y}_1(i) = 0$ when $\hat{\pi}_1(i) < \hat{\pi}_0(i)$. (The notation, used here and in the Appendix, $\pi_k(i), k = 0, 1$ instead of π_k is due to the presence of explanatory variables in the binary component.) The percentage of zero and nonzero values correctly identified is 59.38 (58.36 for the normal model). In particular, the model underestimates the number of zero values in the data (as only 35.1% of the zeroes are correctly identified). This can be due mainly to the absence among the measured covariates of variables describing personal and psychological factors that can influence the decision to consume alcohol. Limiting our investigation to the 79.1% of nonzero values correctly identified, Fig. 3 reports the Healy's plot [10] of the residuals against either the normal (left panel) or the Skew-Normal (right panel) distribution. The figure exhibits a moderate improvement of the second model.

5 Conclusions

In this paper, the possibility of adopting skew distributions for modeling semicontinuous data is analyzed. In particular, the Skew-Normal and the Skew t distributions are contemplated. The resulting classes of models are useful whenever a nonnegative continuous variable shows an anomalous clump in the zero value and asymmetry and/or kurtosis on positive values. This situation is quite common in many fields of application and can be easily carried out visually through a careful analysis of data histogram. For this particular type of data, the proposed models avoid the use of Box-Cox transformation that, when the clump at zero is too high, does not lead to acceptable results.

The semicontinuous regression models presented here can also be viewed as nonstandard mixture models, being a mixture of two components, one binary and the other with positive support. Moreover, the clump at zero is supposed to have two possible sources: one is due to a binary latent component, and the other to a continuous latent component, whereby this takes negative values. This twofold source of zero values is appealing, for instance, in the study of alcohol expenditure from the IHBS, as presented in the paper. In fact, it allows to account for people having by chance no alcohol expenditure in the short period of observation. For this data set we adopted a probit/Skew-Normal model, which showed to fit better than the probit/normal model.

The results obtained show some evidence that the use of alcohol is related to that of tobacco, both in terms of drinking/not drinking (binary component) and in terms of levels (continuous component). Moreover, a white collar job position seems to reduce the probability of drinking alcohol (binary component), but does not affect the continuous component. The main negative effect on the level of alcohol expenditure seems to be the presence in the family of children aged less than fourteen. Finally, income positively affects alcohol expenditure in both components. In terms of model fitting, the proposed model seems to underestimate the number of zero values, probably due to some missing explanatory variables on the binary part.

Estimates are here obtained by maximum likelihood, via simulated annealing. As these models are in fact nonstandard mixture models, estimates could be alternatively obtained using the EM algorithm. The R functions for the likelihood function of the proposal are available upon request from the authors.

Appendix

Under the assumption that Y_2^* is Skew-Normal, the likelihood function for unit i is:

$$\mathscr{L}_i = \left[\pi_0(i) + \pi_1(i) P(Y_2^*(i) \leq 0)\right]^{I[y_i=0]} \cdot \left[\pi_1(i) f_{\text{SN}}(y_i; \mu_i, \sigma, \alpha)\right]^{I[y_i>0]}$$

$$= \left[\pi_0(i) + \pi_1(i) F_{\text{SN}}(0; \mu_i, \sigma, \alpha)\right]^{I[y_i=0]} \cdot \left[\pi_1(i) f_{\text{SN}}(y_i; \mu_i, \sigma, \alpha)\right]^{I[y_i>0]}$$

$$= \left[\pi_0(i) + \pi_1(i) \left(1 - 2\Phi_2(0, \tfrac{-\mu_i}{\sigma}; -\rho)\right)\right]^{I[y_i=0]} \cdot$$

$$\cdot \left[\pi_1(i) \tfrac{2}{\sigma}\phi(\tfrac{y_i-\mu_i}{\sigma})\Phi\left(\tfrac{\alpha}{\sigma}(y_i - \mu_i)\right)\right]^{I[y_i>0]}.$$

Here $\Phi_2(\cdot; -\rho)$ denotes the distribution function of a standard bivariate normal random variable with $-\rho$ as correlation coefficient, where $\rho = \frac{\alpha}{\sqrt{1+\alpha^2}}$. The log likelihood for unit i is then

$$\ell_i = I[y_i = 0] \cdot \log \left[\pi_0(i) + \pi_1(i) F_{SN}(0; \mu_i, \sigma, \alpha) \right]$$

$$+ I[y_i > 0] \log \left[\pi_1(i) f_{SN}(y_i; \mu_i, \sigma, \alpha) \right].$$

Let us denote:

$$z_i = \frac{y_i - \mu_i}{\sigma}$$

$$u_i = -\frac{\mu_i}{\sigma}$$

$$w(\alpha z_i) = \frac{\phi(\alpha z_i)}{\Phi(\alpha z_i)}$$

$$A_i = \frac{1}{\pi_0 + \pi_1 F_{SN}(0, \mu_i, \sigma, \alpha)}.$$

Then, the score functions for unit i with respect to (some of) the parameters of the Skew-Normal distribution are:

$$S(\mu_i) = I[y_i = 0] \cdot A_i \cdot \frac{\partial}{\partial \mu_i} F_{SN}(0, \mu_i, \sigma, \alpha) + I[y_i > 0] \cdot \frac{1}{f_{SN}(y_i, \mu_i, \sigma, \alpha)} \cdot \frac{\partial}{\partial \mu_i} f_{SN}(y_i, \mu_i, \sigma, \alpha)$$

$$S(\alpha_i) = I[y_i = 0] \cdot A_i \cdot \frac{\partial}{\partial \alpha_i} F_{SN}(0, \mu_i, \sigma, \alpha) + I[y_i > 0] \cdot \frac{1}{f_{SN}(y_i, \mu_i, \sigma, \alpha)} \cdot \frac{\partial}{\partial \alpha_i} f_{SN}(y_i, \mu_i, \sigma, \alpha)$$

where

$$\frac{\partial}{\partial \mu_i} F_{SN}(0, \mu_i, \sigma, \alpha) = \frac{\phi(u_i)\Phi(\alpha u_i)}{\sigma \Phi_2(0, u_i)} \qquad \frac{\partial}{\partial \mu_i} f_{SN}(y_i, \mu_i, \sigma, \alpha) = \frac{z_i}{\sigma} - \frac{\alpha}{\sigma} w(\alpha z_i)$$

$$\frac{\partial}{\partial \alpha} F_{SN}(0, \mu_i, \sigma, \alpha) = \frac{\phi(\sqrt{1+\alpha^2}u_i)}{(1+\alpha^2)\sqrt{2\pi}\Phi_2(0, u_i)} \qquad \frac{\partial}{\partial \alpha} f_{SN}(y_i, \mu_i, \sigma, \alpha) = z_i w(\alpha z_i).$$

We can verify that when $\alpha = 0$, then

$$S(\alpha)_{|\alpha=0} = \sigma \sqrt{\frac{2}{\pi}} S(\mu)_{|\alpha=0}.$$

In fact, when $\alpha = 0$, then $A_i = 1/(\pi_0 + \pi_1 \Phi(y_i)) = A_{i|\alpha=0}$ and

$$S(\mu)_{|\alpha=0} = I[y_i = 0] A_{i|\alpha=0} \cdot \frac{\phi(u_i)}{\sigma \Phi(u_i)} + I[y_i > 0] \cdot \frac{z_i}{\phi(y_i)}$$

$$S(\alpha)_{|\alpha=0} = I[y_i = 0] A_{i|\alpha=0} \cdot \frac{\phi(u_i)}{\sigma \Phi(u_i)} \sigma \sqrt{\frac{2}{\pi}} + I[y_i > 0] \cdot \frac{z_i}{\phi(z_i)} \sigma \sqrt{\frac{2}{\pi}}.$$

References

1. Aristei, D., Pieroni, L.: A double-hurdle approach to modelling tobacco consumption in Italy. Appl. Econ. **40**, 2463–2476 (2008)
2. Azzalini, A.: The skew-normal distribution and related multivariate families (with discussion). Scand. J. Stat. **32**, 159–188 (C/R 189–200) (2005)
3. Azzalini, A., Capitanio, A.: Statistical applications of the multivariate skew normal distribution. J. Roy. Stat. Soc.: Ser. B (Stat. Methodol.) **61**, 579–602 (1999)
4. Blundell, R., Meghir, C.: Bivariate alternatives to the Tobit model. J. Econ. **34**, 179–200 (1987)
5. Capobianco, R., Hutton, J., Stanghellini, E.: Modelling censored data with the skew-normal distribution. In: Proceedings of the 25th International Workshop on Statistical Modelling (IWSM2010), Glasgow, UK, 119–122 (2010)
6. Chai, H., Bailey, K.: Use of log-skew-normal distribution in analysis of continuous data with a discrete component at zero. Stat. Med. **27**, 3643–3655 (2008)
7. Couturier, D., Victoria-Feser, M.: Zero-inflated truncated generalized Pareto distribution for the analysis of radio audience data. Ann. Appl. Stat. **4**, 1824–1846 (2010)
8. Cragg, J.: Some statistical models for limited dependent variables with application to the demand for durable goods. Econometrica: J. Economet. Soc. **39**, 829–844 (1971)
9. Duan, N., Manning, W., Morris, C., Newhouse, J.: A comparison of alternative models for the demand for medical care. J. Bus. Econ. Stat. **1**, 115–126 (1983)
10. Healy, M.J.R.: Multivariate normal plotting. Appl. Stat. **17**, 157–161 (1968)
11. Hutton, J., Stanghellini, E.: Modelling bounded health scores with censored skew-normal distributions. Stat. Med. **30**, 368–376 (2011)
12. Jones, A.: A double-hurdle model of cigarette consumption. J. Appl. Economet. **4**, 23–39 (1989)
13. Lambert, D.: Zero-inflated poisson regression, with an application to defects in manufacturing. Technometrics **34**, 1–14 (1992)
14. Min, Y., Agresti, A.: Modeling nonnegative data with clumping at zero: a survey. J. Iranian Stat. Soc. **1**, 7–33 (2002)
15. Moulton, L., Halsey, N.: A mixed gamma model for regression analyses of quantitative assay data. Vaccine **14**, 1154–1158 (1996)
16. Mullahy, J.: Specification and testing of some modified count data models. J. Economet. **33**, 341–365 (1986)
17. Olsen, M., Schafer, J.: A two-part random-effects model for semicontinuous longitudinal data. J. Am. Stat. Assoc. **96**, 730–745 (2001)
18. Powell, J.: Estimation of semiparametric models. In: Engle, R.F., McFadden, D.L. (eds.) Handbook of Econometrics, vol. IV, pp. 2443–521. North Holland, Amsterdam (1994)
19. R Development Core Team: R: A Language and Environment for Statistical Computing. Vienna, Austria: R Foundation for Statistical Computing (2011)
20. Su, L., Tom, B., Farewell, V.: Bias in 2-part mixed models for longitudinal semicontinuous data. Biostatistics **10**, 374 (2009)
21. Tobin, J.: Estimation of relationships for limited dependent variables. Econometrica: J. Economet. Soc. **26**, 24–36 (1958)

Classification of Multivariate Linear-Circular Data with Nonignorable Missing Values

Francesco Lagona and Marco Picone

Abstract A latent-class mixture model is proposed for the unsupervised classification of incomplete multivariate data with mixed linear and circular components. The model allows for nonignorable missing values and integrates circular and normal densities to capture the association between toroidal clusters of circular observations and elliptical clusters of linear observations. Maximum likelihood estimation of the model is facilitated by an EM algorithm that treats unknown class membership and missing values as different sources of incomplete information. The model is exploited on incomplete time series of wind speed and direction and wave height and direction to identify a number of sea regimes.

1 Introduction

Mixture models [16] provide a general approach to classification in multivariate analysis. The joint distribution of the data is approximated by a mixture of tractable multivariate distributions, which represent cluster locations and shapes. The classification problem is solved as a missing value problem, by treating the unknown cluster membership of each observation as a missing value, to be estimated from the data.

Sea conditions are often monitored by taking circular and linear measurements such as wave and wind direction, wind speed, and wave height. Mixture-based clustering of these data is helpful to identify relevant sea regimes, i.e. specific

F. Lagona (✉)
DIPES, Roma Tre University, Rome, Italy
e-mail: lagona@uniroma3.it

M. Picone
ISPRA, Rome, Italy
e-mail: marco.picone@isprambiente.it

M. Grigoletto et al. (eds.), *Complex Models and Computational Methods in Statistics*, 161
Contributions to Statistics, DOI 10.1007/978-88-470-2871-5_13,
© Springer-Verlag Italia 2013

shapes that the distribution of wind and wave data takes under latent environmental conditions.

Mixture-based clustering of marine data is, however, complicated by the con-currence of different supports on which the data are observed. While a pair of wind speed and wave height is a point in the plane, profiles of wind and wave directions are points in a torus, i.e. a surface generated by revolving a circle in three-dimensional space.

Most of the literature on mixture-based classification methods is associated with the analysis of multivariate data whose components share the same support. Linear observations are typically clustered by mixtures of multivariate normal distributions [2]. Multivariate categorical observations are instead typically clustered by using latent-class models that involve mixtures of multinomial distributions [6]. In directional statistics, while mixtures of Kent distributions are popular in the analysis of spherical data [16], toroidal data have been recently modeled by mixtures of bivariate circular densities in bioinformatics [14] and environmetrics [11, 12]. The literature on the unsupervised classification of multivariate data of mixed type is instead limited and relies either on the availability of complete data information or on the assumption of ignorable missing values [7, 11].

In the case of incomplete data information, the mechanism that generates the missing values should be taken in account, in order to obtain efficient MLEs [17]. If the data are ignorably missing, i.e. the probability of not observing a value does not depend on the unobserved value, then efficient MLEs can be found by maximizing the marginal likelihood, obtained by integrating the likelihood function of the complete data with respect to the missing values. Marine databases are often incomplete because of device malfunctioning and ignorability can be assumed if malfunctioning does not depend on the conditions of the sea. If, otherwise, the data are non-ignorably missing, then a missing value is informative of the unobserved value and a more complex likelihood function must be maximized, obtained by specifying the joint distribution of the complete data and the missing pattern. Such joint models can be classified into either "pattern mixture models" or "selection models." A pattern mixture model factors the joint distribution into the conditional distribution of the response given the missingness pattern, and the marginal distribution of missing indicators. A selection model factors the joint distribution into the marginal distribution of the response and the conditional distribution of missing indicators given the response. Because both specifications have advantages and disadvantages, the choice of a specific approach relies on the purpose of the analysis [13]. In classification studies the interest lies in the identification of clusters according to the marginal distribution of the response, and the selection model has the advantage of directly parameterizing the marginal distribution of the response. A number of different selection models have been proposed in the literature, depending on the assumptions on the process that generates the missing values. Shared random effect models [1] are parsimonious selection models where the missingness is dependent on an unobserved latent class, underlying the observed and unobserved response variables. In this paper we propose a discrete shared random effect model for the unsupervised classification of mixed linear and circular data, affected by nonignorable missing values. A suitable E-M algorithm is exploited for maximum-likelihood estimation, by extending the algorithm developed by Hunt and Jorgensen [7] to handle nonignorable missing values.

2 Data

Relevant wind events in the Adriatic Sea are typically generated by the sirocco wind that blows from SE along the major basin axis, and by the bora flow that creates fine-structured jets within the Dinaric Alps on the eastern Adriatic coast. Wind-wave data are traditionally examined by exploiting numerical wind-wave models. These models, well suited for the analysis of ocean waves, are not flexible enough to account for the complex orography of semi-enclosed basins and, as a result, give biased results in Adriatic studies [3]. When numerical wind-wave models are problematic, sea conditions can be alternatively described in terms of representative wave regimes in specific areas, characterized by the probability of occurrence and corresponding to dominant environmental conditions (e.g., wind conditions), acting in the area and during a period of interest [10]. The data normally exploited for this purpose are environmental observations taken by buoys and tide gauges, located within the study area.

The data that motivated this paper are hourly, quadrivariate profiles with two linear and two circular components: wind speed and wave height, wind direction and wave direction. Hourly wave height and direction were taken in the period 01/18/2011–03/09/2011 by a new-generation buoy, located in the Adriatic sea at about 30 km from the coast. Hourly wind speed and direction were obtained from the nearest tide gauge, located at Ancona.

These data were part of a calibration campaign of the GPS device, used by the buoy and the tide gauge for data transmission. Almost one third of the dataset included profiles with at least a missing value. It was empirically noticed that missing data occurred more often during episodes of bad weather conditions than under good conditions. This could indicate a violation of the missing-at-random assumption, because the conditional probability of device malfunctioning, given the observed data, may depend on the value that the device has not transmitted. For example, high-speed wind and high waves might increase the probability of a buoy transmission error. In this paper we accordingly assume that missing values of wave height and wind speed are nonignorable, and, as a result, the contribution of missing patterns to the likelihood may not be ignored, complicating the model-based classification of the data.

Figure 1 displays the scatterplots of the circular and the linear observations, after discarding the incomplete profiles. For simplicity, bivariate circular data are plotted on the plane, although data points are actually in a torus. In particular, points coordinates in the left-hand side plot of the figure indicate hourly directions *from* which the wind blows and the wave travels, respectively. Points coordinates on the

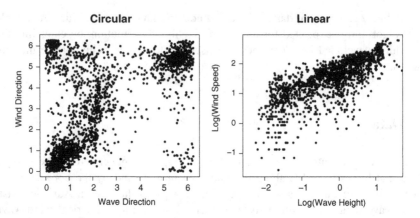

Fig. 1 Hourly observations of wave and wind direction (*left*) and wave log-height and wind log-speed during winter in the Ancona area

right-hand side of the figure indicates the distribution of the log-transformed values of wind speed and wave height. The interpretation of these data is complicated by the complex orography of the Adriatic sea. What we observe, however, could be the result of the mixing of a number of latent regimes of the sea, conditionally to which the distribution of the data takes a shape that is easier to interpret than the shape taken by the marginal distribution. By taking a latent-class approach, we try to identify these latent regimes by associating toroidal and planar clusters that provide an intuitively appealing partitioning of the two scatterplots of Fig. 1, and, when mixed together, adequately approximate the marginal distribution of the data.

3 Model-Based Classification of Linear and Circular Data

The data described in Sect. 2 are gathered in the form of n profiles $z_i = (x_i, y_i, r_i)$, $i = 1, \ldots, n$, which include two circular components, say $x_i = (x_{i1}, x_{i2}) \in (-\pi, \pi]^2$, two linear components, say $y_i = (y_{i1}, y_{i2}) \in \mathbb{R}^2$ and a pattern of missing indicators, say $r_i = (r_{i1}, r_{i2})$, where $r_{i1} = 1$ if y_{i1} is observed and $r_{i1} = 0$ otherwise, while $r_{i2} = 1$ if y_{i2} is observed and $r_{i2} = 0$ otherwise. We model these data by exploiting the mixture

$$f(z|\pi, \beta, \gamma, \eta, \xi) = \sum_{k=1}^{K} \pi_k f_c(x|\beta_k) f_l(y|\gamma_k) p(r_1|y_1, \eta_k) p(r_2|y_2, \xi_k), \quad (1)$$

where $\pi = (\pi_1, \ldots, \pi_K)$ are unknown mixing weights, $\pi_1 + \cdots + \pi_K = 1$, while $f_c(x|\beta_k)$ and $f_l(y|\gamma_k)$ are bivariate densities, respectively, defined on the torus and on the plane, and known up two independent vectors of parameters,

$\beta = (\beta_1, \ldots, \beta_K)$ and $\gamma = (\gamma_1, \ldots, \gamma_K)$. Finally,

$$p(r_1|y_1, \eta_k) = p_{1k}^{r_1}(1 - p_{1k})^{1-r_1}$$
$$p(r_2|y_2, \xi_k) = p_{2k}^{r_2}(1 - p_{2k})^{1-r_2}$$

are two Bernoulli distributions, with parameters linked to the response through a logistic transformation, namely

$$p_{1k} = \frac{\exp(\eta_{0k} + \eta_{1k} y_1)}{1 + \exp(\eta_{0k} + \eta_{1k} y_1)}$$

$$p_{2k} = \frac{\exp(\xi_{0k} + \xi_{1k} y_2)}{1 + \exp(\xi_{0k} + \xi_{1k} y_2)}. \tag{2}$$

Equations (2) are exploited as simple models for the mechanism that generate missing values of wind speed and wave height, within each latent class. In particular, parameters η_{1k} and ξ_{1k} indicate the influence of a value on the probability of not observing that value. When these parameters are equal to zero, the missing mechanism does not depend on the unobserved value and the missing value is ignorable.

In mixture-based classification studies, mixing weights can be conveniently interpreted as the cell probabilities of a latent multinomial vector $w = (w_1, \ldots, w_K)$. As a result, the mixture above can be described as a two-level hierarchical model. At the upper level of the hierarchy, directions (e.g., wind and wave directions) and intensities (e.g., wind speed and wave height) and missing patterns are separately modeled by parametric distributions. These distributions are then non-parametrically associated into K latent classes, at the lower level of the hierarchy. This hierarchy allows to transform the data clustering problem into a missing value problem, where missing class membership w_i of each profile can be predicted by its expectation $\mathbb{E}(w_i|z_i)$, whose kth component is given by

$$\pi_{ik} = \mathbb{E}(w_{ik}|z_i) = \frac{\pi_k f_c(x_i|\beta_k) f_l(y_i|\gamma_k) p(r_{i1}|y_{i1}, \eta_k) p(r_{i2}|y_{i1}, \xi_k)}{\sum_{k=1}^{K} \pi_k f_c(x_i|\beta_k) f_l(y_i|\gamma_k) p(r_{i1}|y_{i1}, \eta_k) p(r_{i2}|y_{i1}, \xi_k)}. \tag{3}$$

The distribution $f_c(x|\beta)$ of bivariate circular data can be specified in a number of different ways [14]. The sine model is a parametric distribution on the torus which embeds naturally the bivariate normal distribution when the range of observations is small. Its density is given by

$$f_c(x; \beta) = \frac{\exp\left(\beta_{11} \cos(x_1 - \beta_1) + \beta_{22} \cos(x_2 - \beta_2) + \beta_{12} \sin(x_1 - \beta_1) \sin(x_2 - \beta_2)\right)}{C(\beta)}, \tag{4}$$

with normalizing constant

$$C(\beta) = 4\pi^2 \sum_{m=0}^{\infty} \binom{2m}{m} \left(\frac{\beta_{12}^2}{4\beta_{11}\beta_{22}}\right)^m I_m(\beta_{11})I_m(\beta_{22}),$$

where

$$I_m(x) = \frac{1}{\pi}\int_0^\pi e^{x\cos t}\cos(mt)dt$$

is the modified Bessel function of order m.

The sine model can be viewed as a bivariate generalization of the von Mises distribution, where β_{12} accounts for the statistical dependence between x_1 and x_2. The two univariate marginal densities

$$f_c(x_i;\beta) = \int_{-\pi}^{\pi} f_c(x;\beta)dx_j = \frac{2\pi}{C(\beta)} I_0(a(x_i)) \exp(\beta_{ii}\cos(x_i - \beta_i)) \quad i = 1,2 \tag{5}$$

depend on the marginal mean angles $\beta_i, i = 1, 2$ and on the shape parameters

$$a(x_i) = \left(\beta_{jj}^2 + \beta_{12}^2\sin^2(x_i - \beta_i)\right)^{1/2} \quad i = 1,2. \tag{6}$$

If $\beta_{12} = 0$, then $a(x_i) = \beta_{jj}$, $i = 1, 2$ and, as a result, x_1 and x_2 are independent and each of them assumes the von Mises distribution with marginal mean angles β_i and marginal concentrations β_{ii}. The conditional distributions

$$f_c(x_i|x_j;\beta) = \frac{f_c(x;\beta)}{f_c(x_j;\beta)} = \frac{\exp(a(x_i)\cos(x_i - \beta_i - b(x_j)))}{2\pi I_0(a(x_i))} \tag{7}$$

are von Mises with conditional mean angles $\beta_i + b(x_j)$ and conditional concentrations $a(x_i)$, where

$$b(x_j) = \arctan\left(\frac{\beta_{12}}{\beta_{jj}}\sin(x_j - \beta_j)\right). \tag{8}$$

In model (1), we use a family of K sine models $f_c(x|\beta_k)$, indexed by the five parameters $\beta_k = (\beta_{1k}, \beta_{2k}, \beta_{11k}, \beta_{22k}, \beta_{12k})$, to define K toroidal clusters centered at (β_{1k}, β_{2k}) and shaped by the parameters $(\beta_{11k}, \beta_{22k}, \beta_{12k})$.

To model the joint distribution of (log-transformed) wind speed and wave height, we use a family of K bivariate normal densities

$$f_l(y;\gamma_k) = N\left(\begin{pmatrix}\gamma_{1k}\\\gamma_{2k}\end{pmatrix}, \begin{pmatrix}\gamma_{11k} & \gamma_{12k}\\\gamma_{12k} & \gamma_{22k}\end{pmatrix}\right). \tag{9}$$

4 Maximum Likelihood Estimation with Non Ignorable Missing Values

Because our data are in the form of incomplete profiles, we, respectively, refer to $x_{i,\text{mis}}$ and $x_{i,\text{obs}}$ as the missing and observed circular components of profile i and, analogously, to $y_{i,\text{mis}}$ and $y_{i,\text{obs}}$ as the missing and observed linear components.

If the missing values are nonignorable, the maximum likelihood estimate of the parameter $\theta = (\pi, \beta, \gamma, \eta, \xi)$ is the maximum point of the marginal log-likelihood function

$$
\begin{aligned}
\log L(\theta) &= \sum_{i=1}^{n} \log \int \sum_{k=1}^{K} \pi_k\, f_c(x_i|\beta_k)\, f_l(y_i|\gamma_k)\, p(r_{i1}|y_{i1}, \eta_k)\, p(r_{i2}|y_{i2}, \xi_k)\mathrm{d}x_{i,\text{mis}}\mathrm{d}y_{i,\text{mis}} \\
&= \sum_{i=1}^{n} \log L_i(\theta) \\
&= \sum_{i=1}^{n} \log \sum_{k=1}^{K} \pi_k L_{ic}(\beta_k) L_{il}(\gamma_k) L_{ir1}(\eta_k) L_{ir2}(\xi_k)) \qquad (10)
\end{aligned}
$$

where $L_i(\theta)$ is the likelihood contribution of the ith profile, $L_{ic}(\beta_k)$ and $L_{il}(\gamma_k)$ are, respectively, the conditional likelihood contributions of the circular and linear components of the ith profile, given the latent class k, and finally $L_{ir1}(\eta_k)$ and $L_{ir2}(\xi_k)$ are the conditional likelihood contributions of the missing patterns, given the latent class k.

Because direct maximization of (10) can be computationally problematic, we exploit an EM algorithm that generates a sequence $(\hat{\theta}_t, t = 1, 2, \ldots)$ of estimates such that $L(\hat{\theta}_t) \geq L(\hat{\theta}_{t-1})$. The algorithm is based on the iterative maximization of the expected value of a complete-data log-likelihood function, computed with respect to the conditional distribution of the unobserved quantities given the observed data. More precisely, we treat the unknown class membership w_i and the unobserved data $(x_{i,\text{mis}}, y_{i,\text{mis}})$ as missing values and define the complete log-likelihood function as a sum of five terms, as follows

$$
\log L_{\text{comp}}(\theta) = \sum_{i=1}^{n} \log L_{i,\text{comp}}(\theta),
$$

where

$$
\begin{aligned}
L_{i,\text{comp}}(\theta) = \sum_{k=1}^{K} w_{ik}\Big\{ &\log \pi_k + \log f_c(x_i; \beta_k) + \log f_l(y_i; \gamma_k) \\
&+ \log p(r_{i1}|y_{i1}; \eta_k) + \log p(r_{i2}|y_{i2}; \xi_k)\Big\}.
\end{aligned}
$$

Given the estimate $\hat{\boldsymbol{\theta}}_t$, provided by the algorithm at step t, a new point $\hat{\boldsymbol{\theta}}_{t+1}$ is computed within step $t + 1$, as follows. We first compute (E step) the expected value of $\log L_{i,\text{comp}}(\boldsymbol{\theta})$ with respect to the conditional distribution of the missing values $(\boldsymbol{w}_i, \boldsymbol{x}_{i,\text{mis}}, \boldsymbol{y}_{i,\text{mis}})$ given the observed data $\boldsymbol{y}_{i,\text{obs}}$, evaluated at $\boldsymbol{\theta} = \hat{\boldsymbol{\theta}}_t$, say

$$\text{(E step)} \qquad Q_i(\boldsymbol{\theta}|\hat{\boldsymbol{\theta}}_t) = \mathbb{E}_t\left(\log L_{i,\text{comp}}(\boldsymbol{\theta})|z_{i,\text{obs}}\right) \quad i = 1, \ldots, n. \tag{11}$$

We then (M step) maximize $Q(\boldsymbol{\theta}|\hat{\boldsymbol{\theta}}_t) = \sum_{i=1}^n Q_i(\boldsymbol{\theta}|\hat{\boldsymbol{\theta}}_t)$ by finding the roots $\hat{\boldsymbol{\theta}}_{t+1}$ of the expected complete data score equations

$$\text{(M step)} \qquad \frac{\partial}{\partial \boldsymbol{\theta}} Q(\boldsymbol{\theta}|\hat{\boldsymbol{\theta}}_t) = \sum_{i=1}^n \frac{\partial}{\partial \boldsymbol{\theta}} Q_i(\boldsymbol{\theta}|\hat{\boldsymbol{\theta}}_t) = \sum_{i=1}^n s_i(\boldsymbol{\theta}|\hat{\boldsymbol{\theta}}_t) = \mathbf{0}, \tag{12}$$

where $s_i(\boldsymbol{\theta}|\hat{\boldsymbol{\theta}}_t)$ is the i-th score vector, obtained by deriving the ith contribution to the expected complete log-likelihood with respect to the parameters.

Variances of the estimates can be found on the diagonal of the inverse of the information matrix $\mathbf{I}(\boldsymbol{\theta})$, which can be consistently estimated by the empirical information matrix

$$\hat{\mathbf{I}} = \sum_{i=1}^n s_i(\hat{\boldsymbol{\theta}}_T)s_i^{\mathsf{T}}(\hat{\boldsymbol{\theta}}_T),$$

where $\hat{\boldsymbol{\theta}}_T$ is the last parameter update, as provided by the algorithm upon convergence.

The practical implementation of both the E- and the M-step of the algorithm is facilitated by the conditional independence assumption between circular and linear data and missing patterns, which holds under (1). As a result, the expected value of the complete log-likelihood function with respect to the conditional distribution of the missing values given the observed data is (at the $(t + 1)$-th step of the algorithm) given by

$$Q(\boldsymbol{\theta}|\hat{\boldsymbol{\theta}}_t) = \sum_{i=1}^n \sum_{k=1}^K \hat{\pi}_{tik} \log \pi_k$$

$$+ \sum_{i=1}^n \sum_{k=1}^K \hat{\pi}_{tik} \mathbb{E}_t\left(\log f_c(\boldsymbol{x}_i; \boldsymbol{\beta}_k)|\boldsymbol{x}_{i,\text{obs}}, w_{ik} = 1\right)$$

$$+ \sum_{i=1}^n \sum_{k=1}^K \hat{\pi}_{tik} \mathbb{E}_t\left(\log f_l(\boldsymbol{y}_i; \boldsymbol{\gamma}_k)|\boldsymbol{y}_{i,\text{obs}}, w_{ik} = 1\right)$$

$$+ \sum_{i=1}^{n} \sum_{k=1}^{K} \hat{\pi}_{tik} \mathbb{E}_t \left(\log p(r_{i1}|y_{i1})|y_{i,\mathrm{obs}}, w_{ik} = 1 \right)$$

$$+ \sum_{i=1}^{n} \sum_{k=1}^{K} \hat{\pi}_{tik} \mathbb{E}_t \left(\log p(r_{i2}|y_{i2})|y_{i,\mathrm{obs}}, w_{ik} = 1 \right)$$

$$= Q_1(\pi|\hat{\theta}_t) + Q_2(\beta|\hat{\theta}_t) + Q_3(\gamma|\hat{\theta}_t) + Q_4(\eta|\hat{\theta}_t) + Q_5(\xi|\hat{\theta}_t). \quad (13)$$

Therefore, the E step of the algorithm essentially reduces to the evaluation of five updating functions. Functions Q_2 and Q_3 can be evaluated by replacing the sufficient statistics by their expected values (see, for example, [12, 18] for the computation of the expected sufficient statistics under bivariate von Mises and normal densities). The expectations in Q_4 and Q_5 can be instead evaluated by computing a Monte Carlo average, obtained by sampling from the conditional distribution of the missing values $y_{i,\mathrm{mis}}$ given the observed values $(y_{i,\mathrm{obs}}, r_i)$. A traditional sampling strategy relies on a Gibbs sampler along with the adaptive rejection algorithm of Gilks and Wild [5], as described in [8]. The M step of the algorithm is carried out by maximizing separately the five updating equations. The first updating function, Q_1, is maximized by

$$\hat{\pi}_{t+1,k} = \frac{\sum_{i=1}^{n} \hat{\pi}_{tik}}{n}, \quad k = 1, \dots, K.$$

Function Q_2 can be maximized by following the iterative procedure suggested in [12], while Q_3 can be maximized by using the well-known updating formulas of the EM algorithm for mixtures of bivariate normal distributions. Finally, maximization of Q_4 and Q_5 reduces to the estimation of a battery of two weighted logistic regressions, each estimated on a augmented dataset, with each $y_{i,\mathrm{mis}}$ filled by a set of m_i samples drawn from the conditional distribution of the missing values given the observed values, each contributing a weight $1/m_i$.

Upon convergence of the algorithm the probabilities $\hat{\pi}_{ik}$ can be exploited to cluster incomplete profiles into K groups by modal allocation, i.e. assigning each profile i to the latent class with the highest probability $\hat{\pi}_{ik}$.

The EM algorithm can get stuck in local maxima of the log-likelihood function or can be attracted by singularities at the edge of the parameter space, where the log-likelihood is unbounded. The presence of multiple local and spurious maxima is well documented in the case of mixtures of heteroschedastic normal distributions [16] and less widely known in the case of bivariate circular distributions [14]. A number of strategies have been proposed to select a local maximizer and detect a spurious maximizer. To avoid local maxima we follow a short-runs strategy (known as the emEM algorithm [4]), by running the EM algorithm from a number of random initializations, stopping at iteration t as soon as

$$\frac{\log L(\hat{\boldsymbol{\theta}}_t) - \log L(\hat{\boldsymbol{\theta}}_{t-1})}{\log L(\hat{\boldsymbol{\theta}}_t) - \log L(\hat{\boldsymbol{\theta}}_0)} \le \tau.$$

We have observed that convergence to spurious maxima is fast (a phenomenon that is well known in the case of mixtures of multivariate normal densities [9]) and can be detected within short EM runs, by monitoring both the class proportions $\hat{\pi}_{tk}$ and the eigenvalues of the covariance matrices

$$\begin{pmatrix} \hat{\beta}_{t11k} & \hat{\beta}_{t12k} \\ \hat{\beta}_{t12k} & \hat{\beta}_{t22k} \end{pmatrix}^{-1} \qquad \begin{pmatrix} \hat{\gamma}_{t11k} & \hat{\gamma}_{t12k} \\ \hat{\gamma}_{t12k} & \hat{\gamma}_{t22k} \end{pmatrix}.$$

After excluding spurious solutions, we select the output of the EM short run that maximizes the log-likelihood, which is then used to initialize a long-run of the EM algorithm.

5   Results

We have estimated a number of mixture models from the data illustrated in Sect. 2, by varying the number of components from 2 to 5. EM short runs were stopped by using a threshold $\eta = 10^{-3}$, typically reached between 10 and 20 iterations, depending on the dimension K of the model. The subsequent long EM run typically required between 100 and 200 iterations to reach convergence (we stopped the algorithm when the log-likelihood difference between successive iterations was less than 10^{-6}). EM short runs were initialized as in [7]. We randomly split the observations into K groups. The first M step is then performed on the basis of these initial groupings. Circular parameters are estimated from the available data by the method of moments, as suggested in [15]. Means and covariance matrices of the normal components are estimated by their empirical counterparts, using the available data, by following [18]. To select the number of components, we computed the Bayesian Information Criterion (BIC) which suggested a model with $K = 2$ components. The model estimates, displayed in Table 1, indicate locations and shapes of three pairs of toroidal and planar clusters, depicted in Fig. 2 through contour lines of bivariate densities.

The first component of the model includes about half of the sample ($\hat{\pi}_1 = 0.46$) and is associated with periods of either calm sea or sirocco episodes. Under this regime, wave and wind directions are weakly correlated. As expected, wind and wave directions are poorly synchronized under good sea conditions, because if wind episodes are weak then wave direction is more influenced by marine currents than by wind direction.

The second component is instead associated with bora episodes. Under the second component, bora jets blowing from north drive high waves that travel along

Table 1 Parameter estimates (standard errors within brackets) of a two-component discrete shared random effect model

Parameter	Component 1	Component 2
β_{1k}	1.29	0.19
(wave mean direction)	(0.07)	(0.03)
β_{2k}	1.34	0.06
(wind mean direction)	(0.05)	(0.06)
β_{11k}	0.77	6.79
(wave directional concentration)	(0.11)	(0.51)
β_{22k}	0.07	4.89
(wind directional concentration)	(0.16)	(0.09)
β_{12k}	0.85	7.65
(wind/wave directional inverse correlation)	(0.23)	(0.28)
γ_{1k}	−0.83	0.33
(wave mean height)	(0.05)	(1.69)
γ_{2k}	1.12	2.00
(wind mean speed)	(0.22)	(0.38)
γ_{11k}	0.64	0.25
(wave height variance)	(0.01)	(0.29)
γ_{22k}	0.41	0.12
(wind speed variance)	(0.29)	(0.64)
γ_{12k}	0.29	0.13
(wind/wave covariance)	(0.03)	(0.10)
η_{0k}	2.85	−3.05
(wind missing mechanism)	(0.00)	(0.02)
η_{1k}	0.01	0.98
(wind missing mechanism)	(0.05)	(0.01)
ξ_{0k}	3.91	−2.1
(wave missing mechanism)	(0.01)	(0.07)
ξ_{1k}	−0.03	1.97
(wave missing mechanism)	(0.01)	(0.01)
π	0.46	0.54
(component weight)	(0.02)	(0.01)

the major axis of the basin. Compared to episodes of calm sea, wind and wave directions appear now strongly synchronized.

This classification is carried out by accounting for possible nonignorable missing values of wave height and wind speed. Interestingly, the estimated coefficients of the two missing mechanisms, η_{12} and ξ_{12}, are significant at a 95% confidence level, indicating that missing values are nonignorable under bora episodes (component 2), which are associated with bad weather conditions. Under good environmental conditions (component 1), such significance vanishes (η_{11} and ξ_{11} are not significant at a 95% confidence level) and the probability to observe a missing value of wind speed and wave height does not depend significantly on the unobserved values of wind speed and wave height.

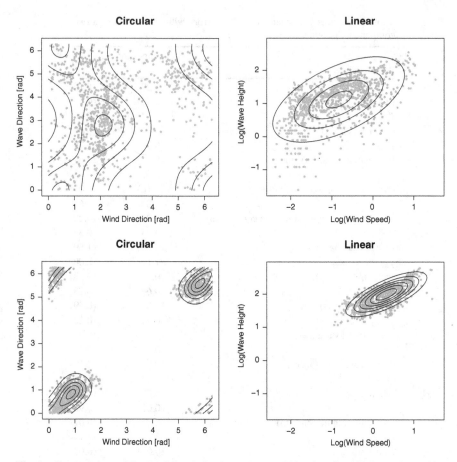

Fig. 2 Contour plots of the conditional circular and normal bivariate densities, as estimated by fitting a 2-component mixture model (first component: top; second component: bottom) and the classification output obtained by modal allocation

Overall, the model indicates that the influence of coastal wind on offshore waves changes under different environmental regimes. The (marginal) weak correlation between wind speed and wave height can be then explained by the presence of a regime under which coastal winds do not generate waves of significant height.

References

1. Albert, P., Follmann, D.: Random effects and latent processes approaches for analyzing binary longitudinal data with missingness: a comparison of approaches using opiate clinical trial data. Stat. Methods Med. Res. **16**(5), 417–439 (2007). DOI 10.1177/0962280206075308. URL http://smm.sagepub.com/content/16/5/417.abstract

2. Banfield, J., Raftery, A.: Model-based Gaussian and non-Gaussian clustering. Biometrics **49**(3), 803–821 (1993). URL http://www.jstor.org/stable/2532201
3. Bertotti, L., Cavalieri, L.: Wind and wave predictions in the Adriatic sea. J. Marine Syst. **78**, S227–S234 (2009)
4. Biernacki, C., Celeux, G., Govaert, G.: Choosing starting values for the EM algorithm for getting the highest likelihood in multivariate Gaussian mixture models. Comput. Stat. Data Anal. **41**(3–4), 561–575 (2003). DOI 10.1016/S0167-9473(02)00163-9. URL http://www.sciencedirect.com/science/article/pii/S0167947302001639
5. Gilks, W.R., Wild, P.: Adaptive rejection sampling for Gibbs sampling. Appl. Stat. **41**, 337–348 (1992)
6. Hagenaars, J., McCutcheon, A.: Applied Latent Class Analysis. Cambridge University Press, Cambridge (2002)
7. Hunt Land Jorgensen, M.: Mixture model clustering for mixed data with missing information. Comput. Stat. Data Anal. **41**(3–4), 429–440 (2003). DOI DOI: 10.1016/S0167-9473(02)00190-1. URL http://www.sciencedirect.com/science/article/B6V8V-472JRC1-13/2/1564c6358976c518c96c320c38dd052e
8. Ibrahim, J.G., Lipsitz, S.R.: Missing covariates in generalized linear models when the missing data mechanism is non-ignorable. J. Roy. Stat. Soc. **B 61**, 173–190 (1999)
9. Ingrassia, S., Rocci, R.: Degeneracy of the EM algorithm for the MLE of multivariate Gaussian mixtures and dynamic constraints. Comput. Stat. Data Anal. **55**, 1715–1725 (2011)
10. Lagona, F., Picone, M.: A latent-class model for clustering incomplete linear and circular data in marine studies. J. Data Sci. **9**, 585–605 (2011)
11. Lagona, F., Picone, M.: Maximum likelihood estimation of bivariate circular hidden Markov models from incomplete data. J. Stat. Comput. Simul. 1–15 (2012) URL http://www.tandfonline.com/doi/abs/10.1080/00949655.2012.656642. DOI 10.1080/00949655.2012.656642
12. Lagona, F., Picone, M.: Model-based clustering of multivariate skew data with circular components and missing values. J. Appl. Stat. **39**, 927–945 (2012). DOI 10.1080/02664763.2011.626850. URL http://www.tandfonline.com/doi/abs/10.1080/02664763.2011.626850
13. Little, R.: Modeling the drop-out mechanism in repeated-measures studies. J. Am. Stat. Assoc. **90**, 1112–1121 (1995)
14. Mardia, K., Taylor, C., Subramaniam, G.: Protein bioinformatics and mixtures of bivariate von Mises distributions for angular data. Biometrics **63**, 505–512 (2007)
15. Mardia, K.V., Hughes, G., Taylor, C.C., Singh, H.: A multivariate von Mises distribution with applications to bioinformatics. Can. J. Stat. **36**(1), 99–109 (2008). URL http://dx.doi.org/10.1002/cjs.5550360110
16. McLachlan, G., Peel, D.: Finite Mixture Models. Wiley, New York (2000)
17. Rubin, D.: Multiple Imputation for Nonresponse in Surveys. Wiley, New York (1987)
18. Shafer, J.: Analysis of incomplete multivariate data. Chapman and Hall, Boca Raton (1997)

Multidimensional Connected Set Detection in Clustering Based on Nonparametric Density Estimation

Giovanna Menardi

Abstract Clustering methods based on nonparametric density estimation hinge on the idea of identifying groups with the level sets of the probability distribution underlying data. Any section of such distribution, at a given threshold, identifies a level set, being the region with density greater than the threshold. The aim is to find the maximum connected components of this region, as the threshold varies. In this way, a hierarchical structure of the number of groups for each threshold is created.

In multidimensional spaces, identification of the connected sets is nontrivial. The use of spatial tessellation such as the Delaunay triangulation has been successfully adopted to this aim but its computational complexity is too high for large dimensions. We discuss the use of an alternative procedure for identifying the connected regions associated with the level sets of a density function. The proposed procedure claims a computational complexity which depends only mildly on the data dimension, thus overcoming the main limitations of the spatial tessellation. The main idea behind this contribution is to emulate the unidimensional procedure to identify connected sets. The method is illustrated with some numerical examples.

1 Introduction

Cluster analysis refers to a general class of methods to explore data with the aim of finding groups of similar observations. The goal is, usually, pursued on the basis of some criterion based on the distance between observations, or alternatively by evaluating the density underlying data. The latter approach, having an explicitly probabilistic motivation, goes back to [8], who defined groups as "regions of high density separated from other such regions by regions of low density." Data are

G. Menardi (✉)
Department of Statistical Sciences, University of Padua, Padua, Italy
e-mail: menardi@stat.unipd.it

M. Grigoletto et al. (eds.), *Complex Models and Computational Methods in Statistics*,
Contributions to Statistics, DOI 10.1007/978-88-470-2871-5_14,
© Springer-Verlag Italia 2013

supposed to be a sample of *i.i.d.* realizations from an unknown probability density function. The estimation of such function allows for detection of the high density regions, which approximate the population clusters.

The actual formalization of this approach may follow either a parametric or a nonparametric route: model-based clustering (*e.g.* [7]) rests on the idea that each cluster corresponds to a subpopulation, typically belonging to some parametric family. The overall population is then modeled as a mixture of these subpopulations. Alternatively, density estimation is performed by nonparametric methods. The two classes of methods differ not only for the approach adopted to estimate the density function, but there is also a conceptual difference. The nonparametric approach links the clusters directly to the modes of the density underlying the data: any section of the density function identifies a density level set, being the region with density above the threshold level. Clusters are associated with the maximum connected regions of the level set. As the threshold varies, so the cluster structure varies and may be represented according to a hierarchical structure in the form of a tree, where each leave corresponds to a mode of the density function. Clustering procedures based on this latter approach claim some potentialities, compared to the distance-based competitors. The correspondence between groups and high density regions makes possible both the detection of clusters having arbitrary shape and the conceptual definition of the number of groups, which is then operatively estimable. Some examples of procedures which follow this approach are [2, 10–12].

A further nonparametric density-based clustering technique has been proposed by [1] and it is referred to as `pdfCluster`. It claims some main original contributions: an automatic procedure for recognizing the number of groups is provided, by enlightening a connection between the number of the modes of f and the number of connected sets associated with different sections of the density function. Moreover, detection of the high-density connected regions is performed by a suitable manipulation of the Delaunay triangulation. The observations with low density are, then, allocated to the detected clusters by following a logic typical of supervised classification. Beyond the mentioned methodological advantages, the application of the procedure has resulted very effective in several real domains. See, for instance, [3].

The convenience of using a clustering approach based on nonparametric density estimation is limited by a main weakness which this work is meant to focus on. While in the univariate setting detection of the connected components is trivial, in multidimensional spaces these are not explicitly defined, and operatively difficult to identify. In particular, the strategy proposed by [1] has a computational complexity which depends only mildly on the sample size, but it grows exponentially with the dimensionality of data, thus making the application of the procedure unfeasible for large dimensions.

In this work an alternative to the Delaunay triangulation is proposed, and illustrated by real data examples. In Sect. 2 the use of the Delaunay triangulation in the context of density-based clustering methods is described. Section 3 presents an alternative idea to tackle the issue of detecting high density connected sets. Some applications to real data are illustrated in Sect. 4.

2 The Use of Delaunay Triangulation to Detect Connected Sets

Let $\mathcal{X} = (x_1, \ldots, x_n)'$, $x_i \in \mathbb{R}^d$ be the observations to be clustered, supposed as a sample of n independent and identically distributed realizations of a d-dimensional random vector x with unknown probability density function f.

For each constant k, $0 \leq k \leq \max f$, the level set $R(k)$ may be defined as

$$R(k) = \{x \in \mathbb{R}^d : f(x) \geq k\}.$$

When f is unimodal, $R(k)$ is a connected region, *i.e.* it cannot be described as the union of two or more nonempty separated sets. Otherwise, it may be connected or not. If it is disconnected, it is formed by two or more connected components, corresponding to the regions around the modes of f which are encountered by the section at the k level. See Fig. 1 for a simple bidimensional illustration.

The number of connected components of $R(k)$ varies with the threshold k, thus generating a hierarchical structure which may be represented in the form of a tree, which associates the number of connected components of $R(k)$ with each possible choice of k [8]. See the right panel of Fig. 1 for an illustration. The empirical analogue of $R(k)$ can be naturally defined as $\hat{R}(k) = \{x \in \mathbb{R}^d : \hat{f}(x) \geq k\}$, with $\hat{f}$ a nonparametric estimate of f.

The definitions of $R(k)$ and its connected components remain conceptually unaltered in any dimension, but the feasibility of their detection decreases when d increases. In the univariate setting, the existence of a total order relation allows for an easy detection of connected sets because of their direct correspondence to intervals. In multidimensional spaces, instead, there is no obvious representation of connected regions, and their detection is, then, nontrivial. A first artifice to simplify the problem is to restrict the attention to the set

$$S(k) = \{x_i \in \mathcal{X} : \hat{f}(x_i) \geq k\}$$

and to the identification of its connected components. The search is thus moved from a continuous multidimensional space to a finite and discrete set. Next, notions from graph theory may be exploited to convert detection of connected components of $S(k)$ into detection of connected components of a graph $\mathcal{G}$ built on it. Clearly, a key matter becomes to suitably set the edges of a graph whose vertices are the elements of $S(k)$.

In pdfCluster, the issue is faced by means of geometric procedures such as the Delaunay graph, being the dual of the Voronoi tessellation. Given a set of data $\mathcal{X} = (x_1, \ldots, x_n)'$, the Voronoi diagram of $\mathcal{X}$ is a partition of $\mathbb{R}^d$ in n regions $V(1), \ldots, V(n)$ such those each region $V(i)$ is the set of all the points closer to x_i than to any other point of $\mathcal{X}$, $i = 1, \ldots, n$. From the Voronoi tessellation, the Delaunay triangulation can be formed by connecting through an edge the pairs of points (x_i, x_j), when the associated regions of the Voronoi tessellation share

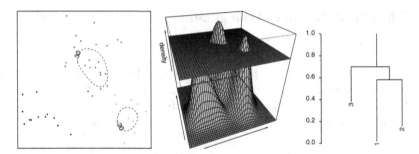

Fig. 1 A sample of size 50 simulated from three subpopulations (*left plot*). A section at the level *k* cuts the trimodal density function (*middle plot*) and identifies two connected components of the level set. The level sets are indicated on the left panel with a *dashed line*. The *right plot* illustrates the cluster tree which associates the number of connected components of $R(k)$ with each choice of *k*

a portion of their boundary facets. These edges partition the space into a set of new polyhedra. Since, in $\mathbb{R}^2$, these polyhedra are triangles this tessellation is often referred to as Delaunay *triangulation*. From a computational point of view the Delaunay triangulation may be obtained directly, without building the Voronoi diagram.

After building the Delaunay graph, the sample points belonging to $S(k)$, and the edges connecting these points only are considered, for each section k of the density estimate $\hat{f}$. The connected components of the subgraphs associated with the $S(k)$'s are then identified as the sets of path-connected observations, namely observations which are pair-wise connected through an edge of the subgraph. This task is straightforward and several algorithms have been designed. See, for example, the depth-first search and the breadth-first search algorithms (see, e.g., [4]). An illustration of the use of the Delaunay triangulation to find high-density connected set is given in Fig. 2 and the main step of the whole pdfCluster procedure is summarized in Table 1.

The use of the Delaunay triangulation is proven to be very suitable to the aim of finding connected sets in multidimensional space: first, the edges of the Delaunay triangulation are the facets of polyhedra in $\mathbb{R}^d$, which are convex sets and, then, connected by definition. Thus, for large n, each polyhedron approximates a corresponding connected component of the unobserved set $R(k)$. Second, the Delaunay triangulation may be considered as a natural generalization of the univariate procedure for detecting the connected components, because spatial tessellation defines a contiguity relationship in the space: sample points are connected by an edge of the Delaunay triangulation when the associated Voronoi regions are contiguous.

The computational complexity to compute the Delaunay triangulation is very competitive with standard methods based on distance measures when $d = 2$ because at most $O(n \log n)$ operations are required. However, the number of required operations has order $O(\frac{n^{\lfloor \frac{d}{2} \rfloor}}{\lfloor \frac{d}{2} \rfloor!})$ for larger d, thus making the triangulation unfeasible for large dimensions.

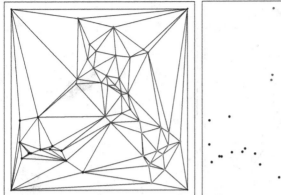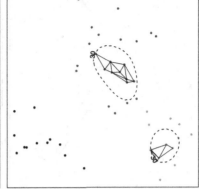

Fig. 2 *Left panel*: Delaunay triangulation of the data in Fig. 1. *Right panel*: polyhedra resulting by removing the vertexes not belonging to $S(k)$ and the edges connecting them from the triangulation, for a given k

Table 1 Main steps of the pdfCluster procedure

---

compute $\hat{f}$;
find the Delaunay triangulation $\mathcal{D}$ of $\mathcal{X}$
while $0 \le k \le \max \hat{f}$ **do**
    identify $S(k) = \{x_i : \hat{f}(x_i) \ge k\}$;
    extract from $\mathcal{D}$ the subgraph $\mathcal{D}_k$, formed by vertices and edges in $S(k)$;
    find the graph connected components of $\mathcal{D}_k$ (e.g. by depth-first-search);
    next k;
end while
build the cluster tree and form the high-density clusters;
allocate remaining points to the high-density cluster for which the log-likelihood ratio
  between the two highest densities, conditional to the group, is maximum.

---

3 Generalizing the Detection of Univariate Connected Sets

In this section, a procedure to approximate the connected components of $S(k)$, alternative to the Delaunay triangulation and computationally feasible even in large dimensions, is presented. The main idea behind this contribution is to emulate the unidimensional procedure to identify connected sets. Subsets of the real line are connected if and only if they are intervals. Hence, an incremental scheme may be adopted to operationally detect connected sets, so that the high-density observations are aggregated in succession, if they result path-connected. As $\mathbb{R}$ is totally ordered, the path between observations is somehow forced, being the segment joining pairs of contiguous observations. Thus, when the path connecting two high-density data points is broken by a low density observation, a new connected set arises.

This way of proceeding is clearly not directly applicable in $\mathbb{R}^d$, because the probability that a (low density) data point lies on the path connecting two

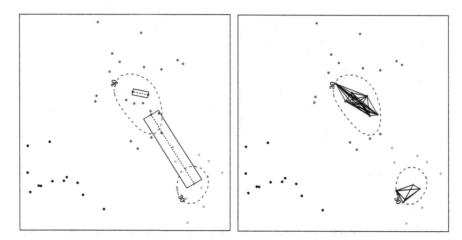

Fig. 3 An illustration of the proposed procedure. On the left, the pair formed by the points with larger distance is not connected because there exist low density points in the path joining them; the other considered pair is path-connected. On the right, the derived subgraph associated with the dashed level set

(high density) observations is zero. However, the unidimensional scheme may be generalized to $\mathbb{R}^d$ by relaxing, in some way, the idea of connecting path between the observations and allowing that such path is d-dimensional itself. Operationally, given a threshold k of the density estimate $\hat{f}$, the level set $S(k)$ is identified. Then, for each pair $(x_i, x_j) \in S(k)$, the two points are said to be connected if the hypercylinder of radius ϵ and the two basis centered at x_i, x_j does not contain any lower density observation. Equivalently, x_i and x_j belong to two distinct connected components of $S(k)$ if at least one lower density point has distance less than ϵ from the line segment joining x_i and x_j. The procedure allows to find the path-connected points of the elements in $S(k)$. The connected components of $S(k)$ may be then easily identified as union of the connected pairs which share at least an observation. An illustration for $d = 2$ is given in Fig. 3.

Except for the procedure to detect high-density connected components, clustering then proceeds by faithfully following the `pdfCluster` approach, as summarized in Table 2.

Clearly, a critical step of the procedure is the choice of the radius of the hypercylinders. Too small values for ϵ would lead to connect observations belonging to distinct connected sets, while too large values would result in the identification of spurious groups. While a choice of ϵ targeted to the specific problem would be desirable, this is a project by itself and is not tackled here. For the time being, intuition, also endorsed by some preliminary simulations, suggests to select the radius of each cylinder as proportional to the distance between the observations it links, with constant less than 1. When groups do not overlap, this choice turns out to be robust to a wide range of choices for the constant of proportionality.

Table 2 Main steps of the method for detecting high-density connected sets

---

compute $\hat{f}$;
while $0 \leq k \leq \max \hat{f}$ **do**
   identify $S(k) = \{x_i : \hat{f}(x_i) \geq k\}$;
   initialize the graph $\mathcal{G}_k$ with vertices $S(k)$ and no edges;
for all $(x_i, x_j) \in S(k)$ **do**
   find the hypercylinder $H(i, j)$ of radius ϵ and basis centered at x_i, x_j
   if $\exists\, x_k \in \mathcal{X} \backslash S(k)$ such that $x_k \in H(i, j)$ **then**
     add edge (x_i, x_j) in $\mathcal{G}_k$;
  end if
  end for
   find the graph connected components of $\mathcal{G}_k$ (e.g. by depth-first-search);
   next k;
end while
build the cluster tree and form the high-density clusters;
allocate remaining points to the high-density cluster for which the log-likelihood ratio
  between the two highest densities, conditional to the group, is maximum.

From a computational point of view, $O(n)$ comparisons are performed to evaluate the path-connected elements of each data point. Moreover, $O(n)$ operations are required by the "depth-first search" algorithm to identify the connected components of $S(k)$. These operations are executed for a grid of considered values of k. Although strongly dependent on the sample size, the advantage of the procedure is that it is independent of d.

4 Numerical Examples

An application of the proposed procedure to several data sets has been performed with the twofold aim of comparing the detected partitions with the clusters found by running alternative methods, and testing the ability of the proposed method in reconstructing an existent clustering structure (this reason explains the use of data with a known label class). As competing methods, the original pdfCluster method and a hierarchical distance-based method with complete linkage have been considered.

For the actual implementation of the proposed procedure, a kernel density estimator with Gaussian kernel and diagonal smoothing matrix has been selected. The smoothing parameters have been set to the asymptotically optimal value under the assumption of multivariate normality. While this choice may appear naive in a context where the presence of groups is expected and the assumption of normality is then likely not to hold, it produces sensible results in most of the applications. Concerning the radius of the hypercylinders, the following examples refer to setting the parameter as proportional to the distance between the observations it links, with constant equal to $1/5$. The number of groups is automatically selected when pdfCluster and the proposed procedure are run. Instead, for the distance-based method the number of clusters has been set to the actual number of groups in data.

In order to allow for comparison with pdfCluster, we consider here data with moderate dimension only or we operate a dimension reduction pre-processing step by means of principal components analysis. It is worth to remind, however, that the computational complexity of the proposed procedure to find the connected high-density regions does not change with the data dimension.

The first considered example is known as the *Flea-beetles* data set [9] and it includes 74 measures of 6 body characteristics from three species of beetles: Ch. concinna, Ch. heptapotamica, and Ch. heikertingeri, corresponding to the groups we aim at identifying.

As a second example, we cluster the *Wine* data [6], including 13 chemical characteristics of 178 wines derived from three different cultivars (Barolo, Grignolino, Barbera), which we aim at reconstructing. Clustering has been performed on the first three principal components of the data.

The third considered data set, referred to as the *Olive oil* data [5], represents 8 chemical measurements on $n = 572$ specimens of olive oil produced in three macro-areas of Italy: South, Sardinia island, Centre-North. These macro-areas result from the aggregation of 9 sub-regions of Italy: Apulia North and South, Calabria, Sicily, Sardinia, coast and inland, Umbria, Liguria East and West. The clustering algorithms have been applied to reconstruct the geographical origin of the specimens. Since the raw data are of compositional nature, totalling 10,000, an additive log-ratio transform has been adopted, following [1]. Clustering has been performed on the first five principal components of the data.

Clusters found on the *Flea-beetles* data (see Table 3) by means of the proposed procedure perfectly overlap the actual partition of data in the three species. The same partition is obtained by running the original pdfCluster method while the hierarchical procedure misclassifies a few units.

Concerning the *wine* data (Table 4), the procedure is able to reconstruct the actual clustering structure with an accuracy greater than 95%. Notably, it again produces the same partition as the one obtained by applying the original pdfCluster method with the Delaunay triangulation. Groups obtained with the distance-based method, instead, do not match well the actual cultivars of origins of the wines.

Clustering the *olive oil* data (Table 5) turns out to be more challenging than the other applications because groups present a high degree of overlapping. The distance-based method produces a partition which does not even closely match the actual groups. The proposed strategy identifies four groups, instead of the three geographical macro-area of origin of the oils, because the observations originated from the Centre-North macro-area, are splitted into two groups. Except for the additional cluster, allocation of the observations closely corresponds to the one produced by using the Delaunay triangulation. However, the forth detected group is still very homogeneous, gathering observations which come exclusively from the Northern regions. Moreover, a suitable setting of the ϵ parameter entails a greater agreement with the partition produced by the original pdfCluster method. At

Table 3 *Flea beetles* data: cross-classification frequencies of the actual groups and of the results of different clustering methods

	New procedure			PdfCluster			Complete linkage		
	1	2	3	1	2	3	1	2	3
Ch. concinna	21	0	0	21	0	0	16	5	0
Ch. heptapotamica	0	22	0	0	22	0	0	22	0
Ch. heikertingeri	0	0	31	0	0	31	0	0	31

Table 4 *Wine* data: cross-classification frequencies of the actual groups and of the results of different clustering methods

	New procedure			PdfCluster			Complete linkage		
	1	2	3	1	2	3	1	2	3
Barolo	58	0	1	58	0	1	59	0	0
Grignolino	2	5	64	2	5	64	47	21	3
Barbera	0	48	0	0	48	0	0	1	47

Table 5 *Olive oil* data: cross-classification frequencies of the actual groups and of the results of different clustering methods

	New procedure				PdfCluster			Complete linkage		
	1	2	3	4	1	2	3	1	2	3
Apulia.north	25	0	0	0	2	0	0	25	0	0
Apulia.south	206	0	0	0	206	0	0	3	203	0
Calabria	56	0	0	0	56	0	0	53	3	0
Sicily	36	0	0	0	30	0	6	26	10	0
Sardinia.inland	0	65	0	0	0	65	0	65	0	0
Sardinia.coast	0	33	0	0	0	33	0	33	0	0
Liguria.east	0	0	49	1	0	1	49	37	0	13
Liguria.west	0	8	9	33	0	41	9	0	0	50
Umbria	0	0	51	0	0	0	51	51	0	0

this regard, Fig. 4 shows the number of clusters detected by the proposed procedure as a function of the radius ϵ: it is reassuring to note that the clustering solutions showing the greatest stability to different values of ϵ are also the ones with three and four groups.

5 Further Remarks

The nonparametric approach to the clustering problem, in a density-based framework, relies on a common definition of groups as the maximum connected components associated with the level sets of the density underlying data. Assumed this model as a basis of the clustering procedures, these are characterized by different

Fig. 4 Number of clusters detected in the *olive oil* data as the radius ϵ varies. The radius of each hypercylinder is computed as proportional to the distance between each pair of observations. The *x*-axis reports the constant of proportionality

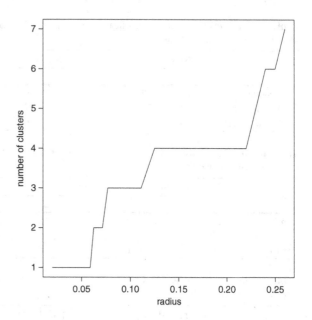

ways of approximating the unknown involved features, namely (1) the level sets of the density function and (2) their associated connected components. While leaving unchanged the estimation of the former, with respect to the method of [1], in this work a procedure alternative to the Delaunay triangulation has been presented, pursuing the same task of approximating the latter unknown component.

In addition to pursuing the same task, the proposed procedure and the Delaunay tessellation share a further conceptual analogy, being both interpretable as a possible generalization of the univariate procedure: consider, in one dimension, two observations which belong to the same level set and are contiguous. This is equivalent to say that there are no lower density points in the path (segment) connecting the two observations. Then, the two observations belong to the same connected set. The Delaunay triangulation translates this idea by extending the concept of contiguity in d dimensions. The proposed procedure, instead, extends the concept of path connecting the observations.

Beyond these analogies, the two procedures cannot, in general, guarantee the same results. However, when the number of connected components is estimated to be the same by the two procedures, also the composition of the high-density clusters is likely to mostly overlap, since the two procedures are based on the same estimate of the level sets; in that case, also the (posterior) probabilities of cluster membership will be comparable in the two procedures, because proportional to the estimated density conditional to the groups. Since the number of detected connected components depends on ϵ, different setting of this parameter may determine a larger or smaller agreement between the partitions produced by using the two procedures.

Then, the proposed procedure turns out to be a valid alternative to the Delaunay triangulation to approximate the connected components of the density level sets. Unlike the Delaunay triangulation, requiring a number of operations which grows exponentially with the data dimensions and depends only mildly on the sample size, the proposed procedure has computational complexity which grows quadratically with the sample size and depends on mildly on the data dimensions. These considerations help us to choose when the use of one procedure is preferable to the other: while large samples having low dimensionality may be efficiently handled by the original `pdfCluster` procedure, higher dimensional data are better to be treated by using the proposed procedure. A rule of thumb would suggest to use the Delaunay triangulation for dimensionality not greater than 6.

The application of the proposed procedure to simulated and real data has shown satisfactory performance because it provides a fair approximation of the original `pdfCluster` method it aims at extending and, in general, produces meaningful partitions of data. While an automatic criterion to select the tuning parameter ϵ would be certainly desirable, it should be borne in mind that some degree of subjectiveness is unavoidable in clustering real data. However, the empirical analysis has shown that sensible results are quite robust to wide ranges of choices of the parameter. As a consequence, the user is suggested to compare results obtained by varying ϵ and to choose the partition which shows the greatest stability. There is room for improvement and looking for an optimality criterion to choose the parameter ϵ as well as alternative and computationally more efficient routes to identify the connected components will be the focus of future research.

Acknowledgements The author wishes to thank prof. Adelchi Azzalini for the fruitful discussions about this work.

References

1. Azzalini, A., Torelli, N.: Clustering via nonparametric density estimation. Stat. Comput. **17**, 71–80 (2007)
2. Cuevas, A., Febrero, M., Fraiman, R.: Cluster analysis: a further approach based on density estimation. Comput. Stat. Data Anal. **36**, 441–459 (2001)
3. De Bin, R., Risso, D.: A novel approach to the clustering of microarray data via nonparametric density estimation. BMC Bioinform. **12**, 49 (2011)
4. Even, S.: Graph Algorithms, 2nd edn. Cambridge University Press, New York (2011)
5. Forina, M., Armanino, C., Lanteri, S., Tiscornia, E.: Food Research and Data Analysis, chapter Classification of Olive Oils from their Fatty Acid Composition, pp. 189–214. Applied Science Publishers, London (1983)
6. Forina, M., Armanino, C., Castino, M., Ubigli, M.: Multivariate data analysis as a discriminating method of the origin of wines. Vitis **25**, 189–201 (1986)
7. Fraley, C., Raftery, A.E.: Model-based clustering, discriminant analysis and density estimation. J. Am. Stat. Assoc. **97**, 611–631 (2002)
8. Hartigan, J.A.: Clustering Algorithms. Wiley, New York (1975)
9. Lubischew, A.A.: On the use of discriminant analysis in taxonomy. Biometrics **18**, 455–477 (1962).

Using Integrated Nested Laplace Approximations for Modelling Spatial Healthcare Utilization

Monica Musio, Erik-A. Sauleau, and Valentina Mameli

Abstract In this work we propose different spatial models to study hospital recruitment, including some potentially explanatory variables, using data from the hospital of Mulhouse a town located in the north-east of France. Interest is on the distribution over geographical units of the number of patients living in this geographical unit. Models considered are within the framework of Bayesian latent Gaussian models. Our response variable is assumed to follow a binomial distribution, with logit link, whose parameters are the population in the geographical unit and the corresponding risk. The structured additive predictor accounts for effects of various covariates in an additive way, including smoothing functions of the covariates (for example a spatial effect). To approximate posterior marginals, which are not available in closed form, we use integrated nested Laplace approximations (INLA), recently proposed for approximate Bayesian inference in latent Gaussian models. INLA has the advantage of giving very accurate approximations and being faster than MCMC methods when the number of hyperparameters does not exceed 6 (as in our case). Model comparison is performed using the Deviance Information Criterion.

M. Musio (✉)
Department of Mathematics and Computer Science, University of Cagliari, Cagliari, Italy
e-mail: mmusio@unica.it

E.-A. Sauleau
Department of Biostatistics, University of Strasbourg, Strasbourg, France
e-mail: ea.sauleau@unistra.fr

V. Mameli
Department of Statistical Science, University of Padua, Padua, Italy
e-mail: valentina.mameli@unipd.it

M. Grigoletto et al. (eds.), *Complex Models and Computational Methods in Statistics*,
Contributions to Statistics, DOI 10.1007/978-88-470-2871-5_15,
© Springer-Verlag Italia 2013

1 Introduction

Analysis of spatial hospital utilization patterns is a fundamental requirement for effective health services planning and hospital management. Moreover, understanding the factors that influence hospital utilization is helpful for identifying reasons for differences in utilization and for formulating policies and programs that promote cost-effective care. Our aim is to examine spatial recruitment in Haute Alsace, a region in the north-east of France, using data from the public hospital of Mulhouse, the biggest town of the region. Several alternative explanatory variables are included in the study. Our analysis focused on global recruitment and not on specific pathologies. Indeed the recruitment is different according to the pathologies and according to elective vs urgent in-patients. Furthermore, the difference between urgent and elective patients is not simply whether they are admitted to an emergency ward or not. There are also urgent in-patients in "classical" wards. In France, there is no routine gathering of this kind of information.

The study of the recruitment requires that we have a geographic reference of patient residence. The address of each patient is reported at the level of a geographical unit (the finest level is the municipality), in which a population at risk can be determined. Our interest is on the distribution over geographical units of the number of patients living in this geographical unit i, say y_i, where the population N_i in the same unit can be considered as the number of persons "at risk" to visit a healthcare provider. We assume that the response variable y_i independently follows a binomial distribution whose parameters are the population N_i and a particular risk per unit p_i. If $\text{logit}(p_i) = \eta_i$, we then have $p_i = \frac{e^{\eta_i}}{1+e^{\eta_i}}$. The effects of covariates of different type are modelled through a geoadditive or structured additive predictor [6, 10], extending the usual linear predictor by adding nonparametric functions for possibly nonlinear effects of continuous covariates and spatial effects. We model the predictor for the geographical unit i in the following way:

$$\eta_i = \mu + f_1(u_{1i}) + f_2(u_{2i}) + \cdots + f_p(u_{pi}) + f_i^{(S)} + \mathbf{z}'\boldsymbol{\beta} + \epsilon_i.$$

Here $f_1, \ldots, f_p$ are possibly nonlinear functions of the continuous covariates $\mathbf{u} = (u_1, \ldots, u_p)'$ and $f^{(S)}$ is a structured spatial effect. The term $\mathbf{z}'\boldsymbol{\beta}$ corresponds to linear effects of (usually categorical) covariates $\mathbf{z}$ and ϵ denotes a spatially unstructured random term. This class of models can be complex and hierarchical, involving fixed and random effects and is particularly suited to Bayesian inference [2, 8].

In a Bayesian framework, spatial models for areal data $f^{(S)}$, based on aggregated disease counts in municipalities, commonly employ Gaussian Markov random field models, in particular the conditional specification known as the intrinsic conditional autoregressive (ICAR) model [4, 15], which assumes that, conditionally on the spatial effects in adjacent geographic units, the effect in a unit follows a normal distribution. The mean of this distribution is the average of the spatial effects in the surrounding units and its precision is proportional to the number of neighbors of this unit. If $f_i^{(S)}$ is the effect in the unit i and $\mathbf{f}_{-i}^{(S)}$ the effects in units other than i in the study area, then the ICAR can be written:

$$f_i^{(S)}|\mathbf{f}_{-i}^{(S)}, \tau^{(S)} \sim \mathcal{N}\left(\frac{1}{n_i}\sum_{j\in\partial_i} f_j^{(S)}, n_i\tau^{(S)}\right).$$

In this formula, n_i denotes the number of units adjacent to each i, ∂_i indicates the set of all of these adjacent units. As common in Bayesian modelling, we specify prior distributions with the precision ($\tau^{(S)}$), the inverse of variance $\left(\tau^{(S)} = \frac{1}{\sigma^2}\right)$. As is often found in disease mapping literature [3,11,15], we have chosen Gaussian priors for $\mathbf{x} = (\mu, f(\cdot), \boldsymbol{\beta}, \epsilon)$, while hyperparameters involved in prior elicitations are not necessarily Gaussian. Our general model is then a latent Gaussian model [15]. To approximate posterior marginals, we use integrated nested Laplace approximations (INLA) [16, 17], recently proposed for approximate Bayesian inference in latent Gaussian models, as an alternative to the Markov Chain Monte Carlo (MCMC) sampling. INLA has the advantage of giving very accurate approximations and being faster than MCMC methods if the number of hyperparameters is not high (the suggested number is less or equal to 6). Implementation of space and space-time models with INLA is presented and explained in detail in [19, 20] and comparisons between INLA and MCMC methods are discussed in an extended form in [7, 9, 17, 22].

The structure of the paper is as follows: in Sect. 2 we describe the data in detail. In Sect. 3 we introduce and justify the model used including details on assumptions on the priors. Results obtained and comparisons of different models are shown in Sect. 4. Concluding discussion is stated in Sect. 5. In the Appendix we describe the basics of the INLA approach and present some pieces of the R code used to implement the models.

2 Data Description and Explanatory Analysis

Data are from the public hospital of Mulhouse (its location is shown on Fig. 1 as "HCP1"), the biggest town of the Haute Alsace region in north-east of France. This region, adjacent to Germany and Switzerland, is 3,525 km^2 and has 756,974 inhabitants (01/01/2010) in a very dense irregular lattice of 377 municipalities ("communes") which are the geographical units we use. The largest distance between the centroids of two geographical units is about 95 km.

In 2009, the hospital of Mulhouse recorded 49,341 sojourns (about 30% of the whole region). Among the patients of these visits, we considered only 33,572 different patients in full-time hospitalization, excluding accesses to the provider for iterative treatment (e.g., dialysis). We have different potential explanatory variables affecting recruitment. More precisely we have to deal with the following requirements:

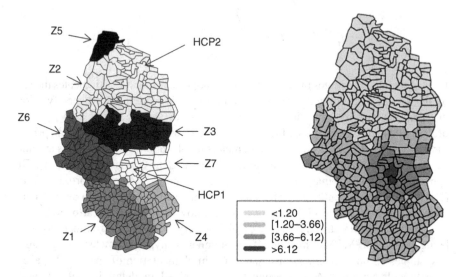

Fig. 1 *On the left*: Proximity zones of the region (Z1: Altkirch, Z2: Colmar, Z3: Guebwiller, Z4: Saint-Louis, Z5: Sélestat, Z6: Thann, Z7: Mulhouse). The studied healthcare provider is HCP1 and the second provider is HCP2. *On the right*: the estimated odds ratio (posterior median) of the spatial effect (ICAR), reference model

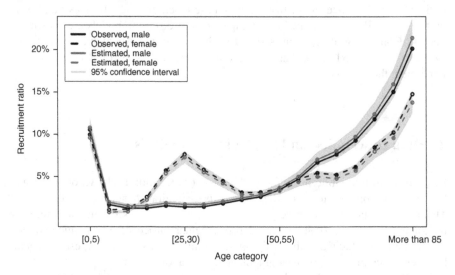

Fig. 2 Observed recruitment ratio per age category and gender, with superimposed the estimated one (anti-logit scale)

(a) Age and gender play a main role on recruitment, highlighted by Fig. 2. The age is categorized in 18 categories from [0,5) up to [80,85) and more than 85 years. Among the 33,572 different patients, 19,109 (56.9%) are female and half of the patients are over 50. In Fig. 2 the high initial points of the curves are explained by births. A peak occurs in young women at the time of deliveries, then gradually as age increases utilization of healthcare becomes more important, especially for men. The obvious role of age, gender, and their interaction is strong enough to justify including these effects in all the models we tested.

(b) Practitioners send their patients preferentially (except for some particular pathologies) to a given healthcare provider. These practitioners recruit patients from several nearby geographical units. This means that the recruitment in a given geographical unit is more "similar" to that in a nearby unit than to that in another random unit in the region. This is the definition of spatial autocorrelation and we can assume that the use of statistical models taking into account the autocorrelation greatly improves the explanation of the recruitment.

(c) The distance or the access time between a healthcare provider and the geographical unit of residence reflects the ease of access to this healthcare provider. The access time may have a greater influence on recruitment in the context of an emergency. Specifically, interest is focused on measuring the attenuation of recruitment with distance (or time), for example by a smoothed function of distance (or time).

(d) A recent French healthcare policy introduced the notion of "proximity zones." The region is divided into several of these zones, shown in Fig. 1 (left side) ("Z1" up to "Z7"), each centered on a healthcare provider to which its patients are recruited. But there are differences between providers according to their technical capacities and competencies. A larger provider also has to recruit patients (for various specific pathologies) from several of these subregions.

(e) Some other covariates can also influence recruitment, such as geographical characteristics or economic status of the geographical units ([1] provides a list of potential covariates that can be included in the study). Herein, we test only two of them:

1. The distance (or the access time) between each geographical unit and a second important healthcare provider ("HCP2" on Fig. 1) assuming that patients living nearer this second provider will prefer to go there rather than to the first.

2. The density of practitioners in each geographical unit (per 1,000 inhabitants).

These considerations lead us to consider Bayesian structured additive regression models [6]. We present the adopted model in detail in the following section.

3 Statistical Model

We assume that the response variable y_{ias}, the number of observed cases living in the ith geographical unit ($i = 1, \ldots, 377$), at age category a ($a = 1, \ldots, 18$) and of sex s ($s \in \{1, 2\}$), follows a binomial distribution with parameters N_{ias} and p_{ias}

where N_{ias} indicates the corresponding population and p_{ias} is the risk. Thus $y_{ias} \sim$ Bin(N_{ias}, p_{ias}). We consider the logit link and the following additive structure for the linear predictor:

$$\text{logit}(p_{ias}) = \eta_{ias} = \mu + \sum_{j=1}^{n_f} f^{(j)}(\mathbf{u}_{ias}) + \sum_{k=1}^{n_\beta} \beta_k z_{kias} + f_i^{(S)} + \epsilon_i. \quad (1)$$

Here, the $f^{(j)}(\cdot)$s are unknown smoothing functions of the covariates in $\mathbf{u}$, the β_ks represent the vector parameters for the linear effect of covariates in $\mathbf{z}$, $f^{(S)}$ is a spatially structured component, and ϵ is a spatially unstructured component. This specification can be considered as a "full" model, including all potential covariates. We assume the following prior distributions:

1. Smoothing functions $f^{(j)}$s are first- or second-order random walk models with precisions $\tau^{(j)}$. A second-order random walk model is commonly used for smoothing data. It is quite flexible due to its invariance to addition of a quadratic trend and it is computationally efficient due to the Markov properties of the joint Gaussian density [5, 12, 15]. Suppose first that $\boldsymbol{u}$ is a time scale or continuous covariate with K equally spaced ordered observations $u_{(1)}, \ldots, u_{(K)}$. Then, if we write $\gamma_k = f(u_{(k)})$, a conditional specification of a first-order random walk for a $\gamma = (\gamma_1, \ldots, \gamma_K)^{\mathrm{T}}$ is $\gamma_k | \gamma_{k-1}, \tau \sim \mathcal{N}(\gamma_{k-1}, \tau)$ where τ is the precision parameter of the random walk. Furthermore $p(\gamma_1) \propto$ const (e.g., a uniform prior is assumed). This prior can be written in a joint specification as $p(\gamma | \tau) \propto \tau^{\frac{N-1}{2}} \exp\left(-\frac{\tau}{2} \sum_2^N (\gamma_k - \gamma_{k-1})^2\right)$. Similarly, a second-order random walk is specified by $\gamma_k | \gamma_{k-1}, \gamma_{k-2}, \tau \sim \mathcal{N}(2\gamma_{k-1} - \gamma_{k-2}, \tau)$ with additional flat priors for γ_1 and γ_2. This improper prior can be written as $p(\gamma | \tau) \propto \tau^{\frac{N-2}{2}} \exp\left(-\frac{\tau}{2} \sum_3^N (\gamma_k - 2\gamma_{k-1} + \gamma_{k-2})^2\right)$. To complete the models, gamma priors are generally assigned to the precision τ. In case of non-equally spaced observations, slight modifications are needed to adjust the error variances (for details, we refer to [5, 12]).

2. The model for the spatial structured component $f^{(S)}$ is an ICAR process, using a first-order neighborhood matrix. With this simple spatial model, one may wonder whether all the spatial effect of the data is taken into account. A way to deal with this question is then to add an unstructured spatial effect, ϵ, for heterogeneity using independent zero-mean Gaussian prior with precision $\tau^{(\epsilon)}$ [4, 13]. The combination of these effects is called "convolution prior" in the disease mapping literature. The unstructured spatial component can be viewed as a proxy for important environmental covariates not gathered and not included in the analysis.

We will assign a $\mathcal{N}(0, \tau^{(\epsilon)})$ for ϵ_i, independent $\Gamma(0.01, 0.01)$ priors to the hyper-parameters $\tau^{(1)}, \ldots, \tau^{(n_f)}, \tau^{(s)}, \tau^{(\epsilon)}$, and a $\mathcal{N}(0, 0.01)$ prior to μ, β_k. We treat these as latent variables. Our models are thus latent Gaussian models, which are a subset of Bayesian additive models with a structured additive predictor, in which Gaussian priors are assigned to μ, all $f^{(\cdot)}$s, $\boldsymbol{\beta}_k$s and ϵs. If we denote by $\mathbf{x} = (\mu, f^{(\cdot)}, \boldsymbol{\beta}, \epsilon)$

the vector of all n Gaussian variables and by $\boldsymbol{\theta} = (\tau^{(1)}, \ldots, \tau^{(n_f)}, \tau^{(s)}, \tau^{(\epsilon)})$ the vector of all hyperparameters, which are not necessarily Gaussian, the posterior marginals of interests can be written as

$$\pi(x_j|\mathbf{y}) = \int \pi(x_j|\boldsymbol{\theta}, \mathbf{y})\pi(\boldsymbol{\theta}|\mathbf{y})\mathrm{d}\boldsymbol{\theta}.$$

In this work we use the integrated nested Laplace approximations to approximate these posteriors. It is a new approach for inference on latent Gaussian models. It substitutes MCMC sampling with a series of numerical approximations, providing very accurate estimates for the posterior marginals in only a fraction of the time needed by MCMC algorithms. In contrast with MCMC, the INLA method does not sample from the posterior. It approximates the posterior with a closed form expression. The approximate posterior marginals can be used to compute summary statistics of interest, such as posterior means, posterior medians, variances or quantiles. Theoretical aspects of the INLA approach are reported in the Appendix. For selecting models, we follow a method in several steps:

1. We first estimate a descriptive model which takes into account age–gender and a spatial effect. The age–gender effect is modelled using a random-walk prior on age for each sex and the spatial effect is estimated using ICAR prior. This model, according to (1), can be written as

$$\mathrm{logit}(p_{ias}) = \mu + I(s = 1)f^{(1)}(u_a) + I(s = 2)f^{(2)}(u_a) + f_i^{(S)}$$

   where I denotes the indicator variable for sex. This model plays the role of a "reference" model.

2. We then add as part of $f^{(j)}$ different potential effects in as many explicative models: distance or access time to the healthcare provider, distance or access time between geographical unit of residence and the second healthcare provider and medical density of the geographical unit. In both cases, the distance is the Euclidian distance between the centroids of two geographical units and the access time is a mean necessary time for reaching destination by road (expressed in minutes). The proximity zones are added using indicator variables (the zone of the healthcare provider under study is used as the reference zone): $\sum_{k=1}^{6} \beta_k I(i \in Z_k)$ where $I(i \in Z_k) = 1$ only if the geographical unit i belongs to the proximity zone Z_k.

3. A multivariate best model is then estimated using previous "significant" effects and the following rules: we consider as significant an effect which yields a better model if it is included than if it is not. The second rule is that only distance or access time is included if both are significant and the most significant is preferred. The goodness of fit of each model is assessed using the Deviance Information Criterion (DIC) [21], a generalization of the Akaike score, which can be computed using INLA [17]. The DIC is defined as $\mathrm{DIC} = \bar{D} + p_D$, decomposed like penalized likelihood measures into two terms: $\bar{D}$ measuring the fit to data and p_D, the effective number of parameters, measuring the complexity

of the models. We also use DIC for comparing models. In [21], the authors suggest that models with DIC values within 1 or 2 of a "best" model are also strongly supported, those with values between 3 and 7 of the "best" are only weakly supported and models with a DIC more than 7 higher than the "best" are substantially inferior.

4. We also check the fit of the best model according to two different processes. The first one involves taking into account the spatial information contained in the data, by comparing the previous best multivariate model with the same model without the ICAR. Moreover, we add an unstructured spatial effect to this best model and search for remaining spatial structure such as autocorrelation or clusters. Secondly, for the covariates whose effects are estimated with a random-walk prior we compared an alternative using linear or quadratic trends.

5. Finally, a sensitivity analysis is carried out to test the robustness of this best model. As stated in [14], a crucial problem in the formulation of Bayesian generalized linear mixed model is the specification of the prior distribution for the random effects precision parameters. For this reason the sensitivity analysis is focused on the precision priors. From the previous best model we have tested different combinations for the parameters of the gamma: 0.01 and 0.01 (mean 1 and precision 0.01), 0.1 and 0.01 (mean 10 and precision 0.001), and finally 1 and 0.01 (mean 100 and precision 0.0001).

The INLA R package is used for implementing models.

4 Results

4.1 Reference Model

Using a first-order random walk prior on age for males and a second-order random walk for females recovers the exact form of these effects on recruitment as shown in Fig. 2. The DIC for the model including these two effects and spatial effect using ICAR is 28,041.10 with an effective number of parameters of 256.48. These values will be used as reference for assessing other models. Figure 1 (right side) displays the exponential of the posterior median of the ICAR spatial effect. This plot has a similar pattern to the observed recruitment ratio (not shown). It shows that there is a low risk of recruitment in the north half-part of the region and a high risk of recruitment in the east and especially around the geographical unit of the healthcare provider.

4.2 Univariate Explicative Models

The first part of Table 1 summarizes the DICs of different models.

The DIC indicates that in addition to age, gender, and a spatial effect, taking into account the practitioner density or the access time to the second provider does not improve the model. Indeed the DIC of these models is less than 2 higher or lower

Table 1 DIC of the different univariate and multivariate explicative models (the best among tested models is emphasized)

Model	p_D	DIC
Reference: age, gender and ICAR	256.48	28,041.10
+ distance to provider	248.97	28,036.25
+ access time to provider	250.49	28,028.58
+ distance to the second provider	224.73	28,030.89
+ access time to the second provider	252.91	28,040.96
+ proximity zone (as factor)	214.51	28,019.32
+ medical density	258.29	28,041.54
+ access time to provider + dist to 2nd provider	201.63	28,010.59
+ access time to provider + prox zone	212.60	28,018.32
+ dist to 2nd provider + prox zone	198.97	28,016.04
+ dist to 2nd provider+ access time to provider + prox zone	197.69	*28,007.01*
+ dist to 2nd provider + access time to provider + prox zone + unstructured spatial effect	198.59	28,007.21

than the DIC of the reference model. On the other hand, taking into account the access time between the geographical unit of patient residence and the healthcare provider greatly improves the model (DIC increases by 12.5). The distance to provider also improves the model but less so than the access time. The distance to the second provider also improves the reference model. This improvement is smaller than that brought by the inclusion of the access time to the first provider and even better than the improvement due to the distance to this provider. Table 1 shows, furthermore, that the proximity zone is the covariate that, among those we tested, improves the initial model the most. We use fixed effects (independent zero-mean Gaussian priors with fixed precisions) and random effects (i.i.d zero-mean Gaussian priors with precision τ, where each τ has a gamma prior with 0.01 and 0.01 parameters) but as the results are similar, only the results with fixed effects are shown.

4.3 Multivariate Explicative Models

Considering that distance and access time to the provider, distance to the second provider and proximity zone improve the DIC with respect to the reference model, these covariates are candidates for selection into the best model in the multivariate step. Due to the strong correlation between the distance and the access time to the provider (Pearson correlation coefficient is 0.86) and due to the fact that the distance improves a little less the DIC, only the access time to the provider is included as potential covariate. The second part of Table 1 shows the DIC of the models adding to the reference model two or three of the tested effects. According to the DIC, the best model among those tested is the model adding the three covariates to the reference model. Its DIC is 28,007 (compared with 28,041 for the reference model).

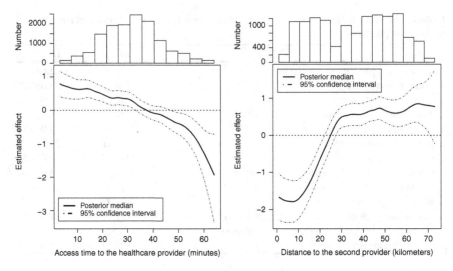

Fig. 3 The access time effect (on the left), and of the distance to second healthcare provider effect (on the right), both on the logit scale

	Odds ratio	95%	Confidence interval
Table 2 Odds-ratios: odds of hospital recruitment in Mulhouse (Z7) divided by odds of hospital recruitment in the other proximity zones with 95% confidence limits			
Z1: Altkirch	1.145	0.901	1.455
Z2: Colmar	3.021	1.725	5.279
Z3: Guebwiller	1.360	0.931	1.987
Z4: Saint-Louis	1.089	0.831	1.425
Z5: Selestat	12.082	3.245	53.178
Z6: Thann	0.908	0.717	1.146

With the same notation as (1), this model can be written as:

$$\text{logit}(p_{ias}) = \mu + I(s = 1)f^{(1)}(u_a) + I(s = 2)f^{(2)}(u_a) + f^{(3)}(u_{2i})$$
$$+ f^{(4)}(u_{3i}) + \sum_{k=1}^{6}\beta_k z_{ki} + f_i^{(S)}$$

where u_a denotes the covariate "age group", while u_2 and u_3 stand, respectively, for "access time to the first provider" and "distance to the second provider."

We now consider the plots of the estimated effects from this best model. Figure 3 (left side) displays the estimated effect of access time to provider plotted against the time. This effect is close to a linear effect on the log odds scale with a slightly decreasing slope for increasing time. Figure 3 (right side) shows the major effect of the second provider when patients live less than 20 km away from it. Finally, using the posterior medians for the proximity zones we calculated odds ratios with 95% confidence intervals. The odds of the recruitment are reported in Table 2. You can see that the odds of recruitment in Mulhouse is about 3 times higher than in Colmar and 12 times higher than in Selestat. Then, even adjusted for the other covariates, a significant under-recruitment persists in the two northern proximity zones (including that of the second healthcare provider).

In this best among tested models, the ICAR process contributes strongly to the fit improvement. Indeed the similar model without ICAR yields a significantly higher DIC: 28,379.86 (compared with 28,007.01) for an effective number of parameters of 101.76 (compared with 197.69). Since the DIC is higher whereas the p_D is lower, the fit to the data is much worse. This implies that covariates alone (access time and distance) do not adequately explain all the spatial correlation in the data. A check for remaining spatial effect has been carried out by a visual inspection of the plot of an unstructured spatial effect added to this best model. The last part of Table 1 allows us to conclude that according to the DIC this new model does not fit the data better than the previous one, since the DIC remains unchanged (28,007) whereas the number of effective parameters increases by 1 (198.59 against 197.69). Furthermore, the plot of the unstructured spatial effect (not shown) reveals no remaining spatial structure but only a seemingly random variation across the region of the different colored categories. In addition, to check the accuracy of the best model on the data we replaced the random-walk by linear or quadratic trend for given covariates and then confirmed that the DICs are worse. From Fig. 3, a linear trend could be a good model for access time to the provider. The estimated slope is decreasing, with median (and mean) -0.0256 and with 95% confidence interval $[-0.0366; -0.01491]$. The DIC of this model is 28,013.96, higher than the DIC for the best model. Also from Fig. 3, a quadratic trend could be a good model for the distance to the second provider. The DIC is then 28,013.36, again worse than that of the best model. These two considerations reinforce the initial option to use smoothing for retrieving accurate estimates of the effects of both these covariates. Finally, sensitivity analysis was performed to check the robustness of the best model to changes of the prior of the precision hyperparameters. All alternative specifications considered yielded very similar results in terms of DIC (not reported) suggesting the robustness of our choices.

5 Discussion

The analysis of healthcare provider recruitment is a fundamental requirement for health services planning. For this we need to select explanatory covariates from potential variables such as distance or access time between providers and the geographical unit of residence of consumers, healthcare organization subregions (like "proximity zones" in France), geographical characteristics or economic status of the geographical units, also taking into account spatial autocorrelation and potential spatial nonstructured effect. Several covariates, such as age and gender, are known to play a role. In this framework of intensive search for a best model or a few relevant models among several complex models, accurate, flexible, and fast methods are needed. Due to the complexity of the situation, Bayesian inference is inescapable and integrated nested Laplace approximations instead of MCMC techniques allow saving time without damaging the results. The application of INLA technique to healthcare provider recruitment seems to be very useful and accurate. Indeed the models are robust to modifications in prior assumptions and their results are globally coherent. The best model among those we tested explains the recruitment in a provider by a spatial autocorrelation, age and gender effects (modelled with two random-walks for age, one for each sex), access time to the provider, distance to a second and concurrent provider and proximity zone. On the other hand, the density of practitioners in each geographical unit does not explain the amount of recruitment. The plot of the ICAR (Fig. 3) shows that autocorrelation is stronger in south part of the region, in proximity zones 1, 4, 6, and mainly 7. This highlights the strong homogeneity of the communes in the zone where the provider stands and a lesser homogeneity in the other ones where the influence of this provider is questioned by the influence of other providers or even of other type of medical care. We have found that access time to the provider and that distance to the second provider play different roles in recruitment, notably because of the highway between the two towns which providers belong to. For example, "communes" located to the south of the provider are not so far in distance but are difficult to reach by roads whereas "communes" to the north, near the second provider, could be far from the provider of interest but are near the highway. This is linked to the presence of a motorway that crosses the region from north to south and links both the cities where hospitals are located. To the north and south of the first healthcare provider, some cities are located relatively far from this provider but are on the motorway. The access time is then a better proxy for recruitment than the distance, which draws a circle around the provider. In contrast, to the north of the first provider, some cities are on the motorway and close enough to this provider but people prefer to go to the second one. In this case, the distance to the second one best explains the recruitment in the first one. We consider here that the number of people at risk is the population of a geographical unit. However, we could also apply to this population a factor representing the proportion of the population that can be recruited, but this "hospitalizability" is different according to the pathology concerned: e.g. 20% of the total population or 30% of men over 75 years, based on the prevalence of diabetes. In [17] are described two useful methods for approximating $\pi(x_i|\theta, \mathbf{y})$ in equation (2) (see Appendix). In this paper we have used one of them, the simplified Laplace approximation.

Appendix

In this section we give an overview of the integrated nested Laplace approximations (INLA). INLA is a method recently proposed for approximating Bayesian inference in structured additive models with latent Gaussian field. It provides a fast alternative to MCMC which is the standard tool for inference in such models. In the following we explain briefly how INLA computes posterior marginal distributions of parameters of interest, for details see Rue et al. [17]. For the sake of discussion, consider the posterior distribution $\pi(\mathbf{x}, \theta|\mathbf{y})$ of a generic Bayesian model, with observation $\mathbf{y}$, latent variable $\mathbf{x}$, and hyperparameters θ.

The INLA approach approximates the posterior marginals of the latent Gaussian field:

$$\pi(x_j|\mathbf{y}) = \int \pi(x_j|\boldsymbol{\theta}, \mathbf{y})\pi(\boldsymbol{\theta}|\mathbf{y})d\boldsymbol{\theta}. \tag{2}$$

This approximation is computed in three steps.

The first step approximates the marginal posterior density $\pi(\boldsymbol{\theta}|\mathbf{y})$ of the hyperparameters $\boldsymbol{\theta}$, $\tilde{\pi}(\boldsymbol{\theta}|\mathbf{y})$, which will be used to integrate out the uncertainty with respect to $\boldsymbol{\theta}$ when approximating the posterior marginal of x_j in the third step. To perform numerical integration it is important to select good evaluation points $\boldsymbol{\theta}_k$ obtained defining a grid of points covering the area of high density for $\log \tilde{\pi}(\boldsymbol{\theta}|\mathbf{y})$.

The second step computes the Laplace approximation or the simplified Laplace approximation of $\pi(x_j|\boldsymbol{\theta}, \mathbf{y})$, for selected values of $\boldsymbol{\theta}$ on the grid.

The third step combines the previous two steps by using numerical integration to obtain the marginal of the latent variables:

$$\tilde{\pi}(x_j|\mathbf{y}) = \sum_k \tilde{\pi}_{\text{SLA}}(x_j|\boldsymbol{\theta}_k, \mathbf{y})\tilde{\pi}(\boldsymbol{\theta}_k|\mathbf{y})\Delta_k,$$

where Δ_k is the area weight assigned to each $\boldsymbol{\theta}_k$, selected in the first step.

A software program, *inla*, implements the INLA techniques. In addition, an interface for the R programming language, which makes the use of the software easier, has been produced and can be downloaded from the web site http://www.r-inla.org. The *inla* program is a useful tool which can solve a wide class of models, including time series models, generalized additive models for longitudinal data, geoadditive models and anova type interaction models. Details and examples for the *inla* program and the INLA library can be found in [18]. The commands for running our "best selected model" are:

```
formula1 = f(time,model="rw2",param=c(0.01,0.01))
+ f(distance,model="rw2",param=c(0.01,0.01))
+ as.factor(zone)
+ f(GU,model="besag",graph=graph,param=c(0,0.01))
+ f(age.male,model="rw1",param=c(0.01,0.01))
+ f(age.female,model="rw2",param=c(0.01,0.01))
model = inla(formula=formula1, family="binomial",
Ntrials=pop, data= dataset, quantiles=c(0.025,0.50,0.975),
control.compute=list(dic=TRUE,mlik=TRUE),
control.predictor=list(compute=TRUE))
```

In the formula command, the f() function is used to specify nonlinear effects in the model. Random walk models of order 1 and 2 are specified inside as model="rw1" and model="rw2", respectively. The ICAR model for spatial effects is defined inside the f() function as model="besag"; this model needs a graph file where the neighborhood structure is specified. In all f() function, param will give values for hyperpriors. The inla() function computes the model marked out in formula. The binomial model is specified by adding family="binomial" in the option list while pop allows to specify the number of trials. The DIC is computed using the command `dic=TRUE`. The command `mlik=TRUE` in `control.compute` gives the marginal likelihood of the model. Using the command `control.predictor` it is possible to compute marginal distributions for each value of the linear predictor. Finally, the retrieved quantiles of posterior distributions are specified by `quantiles()`.

Acknowledgments The first author was supported by the project *start up giovani ricercatori* of the University of Cagliari (Italy). The second author by the *Visiting Professor program* of Regione Autonoma della Sardegna (Italy). The authors thank the Mulhouse hospital for providing the dataset, and the referees for helpful and constructive comments.

References

1. Andersen, R., McCutcheon, A., Aday, L., Chiu, G., Bell, R.: Exploring dimensions of access to medical care. Health Serv. Res. **18**(1), 49–74 (1983)
2. Banerjee, S., Carlin, B., Gelfan, A.: Hierarchical modeling and analysis for spatial data. Monographs on Statistics and Applied Probability, vol. 101. Chapman and Hall/CRC, Boca Raton (2004)
3. Bernardinelli, L., Clayton, D., Montomoli, C.: Bayesian estimates of disease maps: how important are priors? Stat. Med. **14**, 2411–2431 (1995)
4. Besag, J., York, J., Mollié, A.: Bayesian image restoration, with two applications in spatial statistics (with discussion). Annal. Inst. Stat. Math. **43**, 1–59 (1991)
5. Fahrmeir, L., Lang, S.: Bayesian inference for generalized additive mixed models based on Markov random field priors. J. Roy. Stat. Soc. Ser. C: Appl. Stat. **50**(2), 201–220 (2001)
6. Fahrmeir, L., Tutz, G.: Multivariate Statistical Modelling Based on Generalized Linear Models. Springer, Berlin (2001)
7. Fong, Y., Rue, H., Wakefield, J.: Bayesian inference for generalized linear mixed models. Biostatistics **8**(1), 1–27 (2008)
8. Gelman, A., Carlin, J., Stern, H., Rubin, D.: Bayesian Data Analysis. Chapman and Hall, London (1995)
9. Held, L., Schrödle, B., Rue, H.: Posterior and cross-validatory predictive checks: a comparison of MCMC and INLA. In: Kneib, T., Tutz, G. (eds.) Statistical Modelling and Regression Structure – Festschrift in Honour of Ludwig Fahrmeir. Springer, Berlin (2010)
10. Kammann, E., Wand, M.: Geoadditive models. J. Roy. Stat. Soc. Ser. C: Appl. Stat. **52**, 1–18 (2003)
11. Lagazio, C., Dreassi, E., Biggeri, A.: Hierarchical Bayesian model for space-time variation of disease risk. Stat. Model. **1**, 17–29 (2001)
12. Lindgrin, F., Rue, H.: On the second order random walk model for irregular locations. Scand. J. Stat. **35**, 691–700 (2008)
13. Mollié, A.: Bayesian mapping of disease. In: Gilks, W., Richardson, S., Wakefield, J. (eds.) Markov Chain Monte Carlo in Practice, pp. 359–79. Chapman and Hall, New York (1996)
14. Roos, M., Held, L.: Sensitivity analysis in Bayesian generalized linear mixed models for binary data. Bayesian Anal. **6**(2), 231–258 (2011)
15. Rue, H., Held, L.: Gaussian Markov random fields; theory and applications. Monographs on Statistics and Applied Probability, vol. 104. CRC/Chapman and Hall, Boca Raton (2005)
16. Rue, H., Martino, S.: Approximate Bayesian inference for hierarchical Gaussian Markov random field models. J. Stat. Plan. Infer. **137**, 3177–3192 (2007)
17. Rue, H., Martino, S., Chopin, N.: Approximate Bayesian inference for latent Gaussian models by using integrated nested Laplace approximations. J. Roy. Stat. Soc. Ser. B **71**, 319–392 (2009)

18. Martino, S., Rue, H.: Implementing approximate Bayesian inference using integrated nested Laplace approximation: a manual for the *inla* program.   Tech. Rep. 2, Department of Mathematical Sciences, NTNU, Trondheim, Norge (2009)

19. Schrödle, B., Held, L.: A primer on disease mapping and ecological regression using INLA. Comput. Stat. **26**(2), 241–258 (2011). URL http://dx.doi.org/10.1007/s00180-010-0208-2

20. Schrödle, B., Held, L.: Spatio-temporal disease mapping using INLA. Environmetrics **22**(6), 725–734 (2011). DOI 10.1002/env.1065. URL http://dx.doi.org/10.1002/env.1065

21. Spiegelhalter, D., Best, N., Carlin, B., Van der Linde, A.: Bayesian measures of model complexity and fit (with discussion). J. Roy. Stat. Soc. Ser. B **64**, 583–639 (2002)

22. Vanhatalo, J., Pietiläinen, V., Vehtari, A.: Approximate inference for disease mapping with sparse Gaussian processes. Stat. Med. **29**, 1580–1607 (2010)

Supply Function Prediction in Electricity Auctions

Matteo Pelagatti

Abstract In the fast growing literature that addresses the problem of the optimal bidding behaviour of power generation companies that sell energy in electricity auctions, it is always assumed that every firm knows the aggregate supply function of its competitors. Since this information is generally not available, real data have to be substituted by predictions. In this paper we propose two alternative approaches to the problem and apply them to the hourly prediction of the aggregate supply function of the competitors of the main Italian generation company.

1 Introduction

The last 20 years have witnessed in most European and many non-European countries a radical reorganisation of the electricity supply industry. Government-owned monopolies have been replaced by regulated (generally pool) competitive markets, where the match between demand and supply takes place in hourly (in some cases semi-hourly) auctions. The auction mechanism is generally based on a uniform price rule, i.e. once the equilibrium price is determined, all the dispatched producers receive the same price per MWh.

The issue of determining the profit-maximising behaviour of a power company bidding in electricity auctions has been addressed by economists from the both normative (profit optimisation) and positive (market equilibrium) point of views (cf. [1, 5, 13]) and it is faced every hour by the generation companies. Regardless of the bidding strategy a firm pursues, it is necessary to predict the bidding behaviour of its competitors. In particular, each firm has to predict the aggregate quantity offered

M. Pelagatti (✉)
Department of Economics, Quantitative Methods and Business Strategies (DEMS),
University of Milano-Bicocca, Milan, Italy
e-mail: matteo.pelagatti@unimib.it

M. Grigoletto et al. (eds.), *Complex Models and Computational Methods in Statistics,*
Contributions to Statistics, DOI 10.1007/978-88-470-2871-5_16,
© Springer-Verlag Italia 2013

by its competitors for all possible prices before submitting its supply schedule to the market.

In this paper we propose two techniques for forecasting supply functions based on principal component analysis and reduced rank regression (RRR). The techniques are applied to the prediction of the hourly supply functions of the competitors of Enel, the main Italian generation company, as observed in two years of Italian electricity auctions.

Functional time series analysis[1] is a relatively new discipline in the statistical literature, even though the wider-ranging functional data analysis field has a longer history, dating back to the paper of [10] and the works of B.W. Silverman on density function estimation and nonparametric regression. A general framework for the problem of functional time series prediction can be found in [6,7], and the first of the two algorithms proposed in this paper (the one based on principal components) is a special case of the proposal in [7]. Our second algorithm (the one based on RRR), instead, cannot be found in the cited papers and, to the best of our knowledge, has never been explored in the statistical literature.

The paper is organised as follows: Sect. 2 introduces the problem of optimal bidding, Sect. 3 describes the Italian auction rules and the data produced by the market maker, Sect. 4 illustrates the two functional prediction techniques, Sect. 5 applies them to the Italian data and Sect. 6 concludes.

2   Optimal Bidding Behaviour

This sections introduces the problem a generation operator faces in every auction as developed in the economic literature [1, 5, 13]. In order to have interpretable closed-form solutions, economists make a series of simplifying assumptions and approximations that do not seem to reveal significant drawbacks when applied to real data (cf. the applications in the cited papers). However, the functional prediction techniques proposed in this paper can be used also in more involved optimisation problems in which transmission constraints and multi-period profits are taken into account.

If we assume that each firm wishes to maximise its profit in each auction independently from the other auctions (as customary in the literature), then we can summarise the optimisation problem as follows. Suppose that D is the (price-inelastic) demand for electricity, $S_{-i}(p)$ is the aggregate supply function of firm i's competitors for any given price p, $C_i(q)$ is the production cost function of firm i for any given quantity of energy q, then for those values of the residual demand $D - S_{-i}(p)$ that the production capacity of firm i can fulfill, the profit function of firm i is given by

---

[1]Functional time series analysis is the statistical analysis and prediction of sequences of functions. For a rigorous theoretic treatment of the subject, the reader should refer to the book of [2], while the excellent articles of [6, 7] are more operational.

$$\pi_i(p) = p \cdot \left(D - S_{-i}(p)\right) - C_i\left(D - S_{-i}(p)\right). \tag{1}$$

This profit function can be extended to include financial contracts as in [5] or vertical integration (i.e. the situation in which the producer is also a retailer and plays in both sides of the auction) as in [1]. By assuming the continuous differentiability of S_{-i} and C_i, and the concavity of π_i, first order conditions indicate that firm i maximises its profit when he/she offers the quantity $D - S_{-i}(p^*)$ at the price p^* that solves

$$p^* = C_i'\left(D - S_{-i}(p^*)\right) + \frac{D - S_{-i}(p^*)}{S_{-i}'(p^*)}. \tag{2}$$

Now, the quantity D and the supply function S_{-i} are generally unknown, but while D can be predicted using standard time series techniques (e.g. [3, 4, 9] and many articles in the *IEEE Transactions on Power Systems*), the prediction of the function S_{-i} is more involved. The next sections illustrate two techniques for the prediction of such supply functions as observable in auction data.

Notice that the assumption that firms build their optimal bidding strategy by considering each single auction as independent from the other auctions is only an approximation economists need to derive a closed-form solution to the profit maximisation problem. If firms optimise their profit by considering a time-span longer than a single future auction, then the objective function is the actualised sum of many copies of (1), and instead of forecasting a single supply function a sequence of $S_{-i}(\cdot)$ has to be predicted. Even though we do not explicitly consider the case of multiple prediction periods here, the techniques discussed in this paper can be easily extended to that set-up.

3   Auction Rules and Data

According to the rules of the Italian electricity day-ahead market, each production unit can submit up to four "packages" of price-quantity pairs. Each pair contains the information on the quantity (in MWh) a production unit is willing to sell and the relative unitary price (in Euro per MWh). Of course one company usually owns many production units and can, therefore, well approximate its (possibly continuous) optimal supply function using a step function with many steps. All the submitted pairs are sorted by price and the corresponding quantities are cumulated. When the cumulated offered quantity matches the total demand, the system marginal price (SMP) is determined and all the units offering energy up to that price are dispatched. If congestions in the transmission network occur, the national market is split into up to seven (recently reduced to six) zonal markets and the same bids are used to determine new local equilibrium prices. In this case the optimisation problem is more involved than the one discussed in Sect. 2 and the solution has to be found numerically, but predictions of the competitors' zonal supply functions are still necessary.

Table 1 Relevant fields in the Italian electricity auctions database

Producer (seller)	Retailer (buyer)
Operator name	Operator name
Plant name	Unit name
Quantity (MWh) of each offer	Quantity (MWh) of each bid
Price (Euro/MWh) of each offer	Price (Euro/MWh) of each bid
Awarded quantity (MWh) for each offer	Awarded quantity (MWh) for each bid
Awarded price (Euro/MWh) for each offer	Awarded price (Euro/MWh) for each bid
Zone of each offer (plant)	Zone of each bid (unit)
Status of the offer: accepted vs. rejected	Status of the bid: accepted vs. rejected

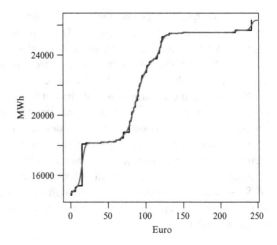

Fig. 1 Supply function of Enel's competitors on 3.12.2008 at 10 a.m. and kernel approximation

Each record of the Italian auction results database[2] (cf. Table 1) contains the price-quantity pair, the name of the offering production unit and the name of the owner of that unit. This allows the construction of the supply function of any firm bidding in the auctions or aggregations thereof (e.g. step function in Fig. 1).

From the above reasoning it is clear that real supply schedules are step functions and, thus, the optimal bidding theory discussed in the previous section is not directly applicable. This issue is generally dealt with by approximating the step functions with continuously differentiable functions obtained though kernel smoothing.[3] Smoothing is also necessary for regularising functions before applying canonical correlation techniques such as RRR (c.f. Sec. 11.5 of [11]). Since supply functions

---

[2]It can be downloaded (on a daily basis) from the market operator web site www.mercatoelettrico. org.

[3]Reference [8] solves the problem of optimal bidding when supply functions are step functions with a given number of steps. However, even in this case the optimal predictions of these step functions need not be step functions, as the prices at which the steps take place may be absolutely continuous random variables and this condition makes the expectation of any random step function a continuous function.

are nondecreasing in price, we use the kernel

$$S(p) = \sum_{k=1}^{K} q_k \Phi\left(\frac{p - p_k}{h}\right),$$

where Φ is the standard normal cumulative probability function, h is the bandwidth parameter and (q_k, p_k) are the observed quantity-price pairs. Notice that the total number of offers K may change in each auction. The derivative of the smoothed function needed in (2) is given by

$$S'(p) = \sum_{k=1}^{K} q_k \frac{1}{h} \phi\left(\frac{p - p_k}{h}\right),$$

with ϕ standard normal density. Figure 1 depicts the actual supply function of Enel's competitors on 3 December 2008 at 10 a.m. and the kernel approximation thereof ($h = 3$ Euro).

4 Supply Functions Prediction

Both prediction techniques proposed in this paper entail some common steps.

The first step consists in sampling the kernel-smoothed function on a grid of abscissa points. This is necessary as the price set on which the function can be evaluated changes in every auction. Since the function can be approximated more accurately where bid pairs are more dense, we sample more frequently in these intervals by using quantiles. In particular, we used 50-iles of unique prices submitted over the entire sample (2007–2008). The forty-nine 50-iles are supplemented with the minimum (0) and the theoretical maximum (500) due to the price capping rule of the Italian market, obtaining 51 time series of ordinate points (quantities). Figure 2 displays one week of Enel's competitors aggregate supply functions sampled at 50-iles. The within-day periodicity and the lower level and slightly different shape of the curve in the weekend are evident from the plot.

The second common step consists in transforming the original ordinate points in a way that preserves the two features of positivity and non-decreasing monotonicity of the original functions also in their predictions. If we denote with $\{p_0, p_1, \ldots, p_{50}\}$ the points in the price grid and with $S_t(p_i)$ the smoothed supply function at time t for price p_i, then we transform the time series as

$$q_{i,t} := \begin{cases} \log S_t(p_i), & \text{for } i = 0; \\ \log\left(S_t(p_i) - S_t(p_{i-1}) + c\right), & \text{for } i = 1, \ldots, 50, \end{cases}$$

where c is a small positive constant that guarantees the existence of the logarithm also in constant tracts of $S_t(p)$ (in our application we set $c = 1$). If we assume that the prediction of $q_{i,t}$, say $\hat{q}_{i,t}$, is unbiased, and the prediction error is approximately

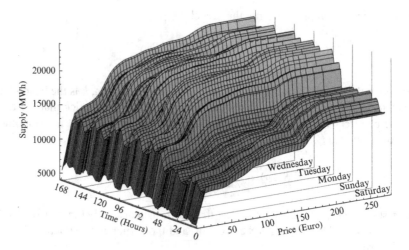

Fig. 2 The supply function sampled at 50-iles over 1 week

normal with standard error $s_{i,t}$, unbiased forecasts of the original function can be recovered as

$$
\hat{S}_t(p_i) = \begin{cases} \exp\left(\hat{q}_{i,t} + s_{i,t}^2/2\right), & \text{for } i = 0; \\ \exp\left(\hat{q}_{i,t} + s_{i,t}^2/2\right) + \hat{S}_t(p_{i-1}) - c, & \text{for } i = 1, \ldots, 50. \end{cases} \tag{3}
$$

Now, since we expect the 51 time series to share some information, it is natural to seek some form of dimension reduction. The two alternative algorithms we propose are based on principal component (PC) analysis and reduced rank regression (RRR). We base the choice of dimension reduction on one month of k-steps-ahead out-of-sample predictions, where $k = 1$ for one-hour-ahead predictions and $k = 24$ for one-day-ahead predictions. In particular, the model is fit to the hourly observations of the years 2007–2008, while the dimension reduction assessment is based on Jan-2009 prediction mean square errors (MSE).

In describing the two algorithms we collect the 51 transformed time series in the vector $\mathbf{q}_t$ and the original supply function ordinate-points in the vector $\mathbf{S_t}$. The predictions are based on lagged responses and deterministic regressors.

Algorithm 1 (Principal component analysis based). *For* $r = \{51, 50, \ldots, 1\}$ *iterate through the following steps.*

1. *Take the first r PCs of $\mathbf{q}_t$ (supply function log increments) based on its in-sample covariance matrix, and name the scores $\mathbf{y}_t$.*
2. *Regress each score $y_{i,t}$ on its lags $\mathbf{x}_{i,t}$ and deterministic regressors $\mathbf{z}_t$ and compute predictions $\hat{\mathbf{y}}_t$.*
3. *Regress the vector $\mathbf{q}_t$ on the predicted scores $\hat{\mathbf{y}}_t$ and a constant.*

4. *Compute the out-of-sample predictions of the supply function $\mathbf{S}_t$ as in (3) and the relative MSE.*

Pick the rank r that minimises the out-of-sample MSE.

In the PC approach, the time series are first reduced in number by taking the best linear approximation to the original data, and then these are predicted using standard time series models. The main advantage of this approach is the freedom left to the analyst to choose the time series model to predict the PC scores. The main drawback is that rank-reduction is not obtained directly for the prediction of future values.

The second approach is based on RRR. Since this technique is less popular than the principal component analysis, we briefly survey its main features. Consider the linear model

$$\underset{n\times 1}{\mathbf{y}_t} = \underset{m\times 1}{\mathbf{C}\,\mathbf{x}_t} + \underset{p\times 1}{\mathbf{D}\,\mathbf{z}_t} + \underset{n\times 1}{\boldsymbol{\varepsilon}_t}$$

where $\mathbf{x}_t$ and $\mathbf{z}_t$ are regressors, $\mathbf{D}$ is a full-rank $n \times p$ coefficient matrix, $\mathbf{C}$ is a $n \times m$ reduced-rank coefficient matrix and $\boldsymbol{\varepsilon}_t$ is a sequence of zero-mean random errors uncorrelated with all the regressors. The fact that $\mathbf{C}$ is reduced-rank means that few linear combinations of the regressors $\mathbf{x}_t$ are sufficient to take account of all the variability of $\mathbf{y}_t$ due to $\mathbf{x}_t$. Now suppose that the rank of $\mathbf{C}$ is $r < \min(m,n)$, then $\mathbf{C}$ can be factorised as $\mathbf{C} = \mathbf{A}\mathbf{B}^\top$, with $\mathbf{A}$ $n \times r$ and $\mathbf{B}$ $m \times r$ matrices. The matrices $\mathbf{A}$ and $\mathbf{B}$ are not uniquely identified, but if one restricts the r column vectors forming $\mathbf{B}$ to be orthonormal, then a least squares solution for $\mathbf{B}$ is found by solving the following eigenvalue problem:

$$\mathbf{S}_{xx|z}\mathbf{V}\Lambda = \mathbf{S}_{xy|z}\mathbf{S}_{yy|x}^{-1}\mathbf{S}_{yx|z}\mathbf{V},$$

where $\mathbf{S}_{ab|c}$ indicates the partial product-moment matrix of $\mathbf{a}$ and $\mathbf{b}$ given $\mathbf{c}$, $\mathbf{V}$ is an orthonormal matrix and Λ is a diagonal matrix. The first r columns of $\mathbf{V}$ provide least square estimates of $\mathbf{B}$. Least squares estimates of $\mathbf{A}$ and $\mathbf{D}$ are found by regressing $\mathbf{y}_t$ simultaneously on $\mathbf{w}_t := \mathbf{B}^\top\mathbf{x}_t$ and $\mathbf{z}_t$. For details on RRR, refer to the excellent monograph [12].

Algorithm 2 (Reduce rank regression based). *For $r = \{51, 50, \ldots, 1\}$ iterate through the following steps.*

1. *Regress the vector $\mathbf{y}_t = \mathbf{q}_t$ on its lags $\mathbf{x}_t$, imposing rank r to the reduced-rank coefficient matrix $\mathbf{C}$, and on the deterministic regressors $\mathbf{z}_t$ without any rank restrictions on $\mathbf{D}$.*
2. *Compute the out-of-sample predictions $\hat{\mathbf{S}}_t$ of the supply function $\mathbf{S}_t$ as in (3) and the relative MSE.*

Pick the rank r that minimises the out-of-sample MSE.

The main advantage of the RRR-based algorithm is that rank-reduction is obtained though the minimisation of the prediction MSE. The drawback is that only (vector) autoregressive models with exogenous variables are allowed.

5 Application to the Italian Electricity Auctions

The two algorithms are applied to the hourly Italian electricity auction results for the years 2007–2008 (17544 auctions); Jan-2009 (744 auctions) is used for determining the rank r as explained in the previous section.

We build models for predicting one-hour-ahead and models for forecasting one-day-ahead. As for the deterministic regressors ($\mathbf{z}_t$) we implement the following three increasing set of variables.

1. Linear trend, $\cos(\omega_j t)$, $\sin(\omega_j t)$, with $\omega_j = 2\pi j/(24 \cdot 365)$ and $j = 1, \ldots, 20$.
2. Regressors at point 1. plus dummies for Saturday, Sunday and Monday.
3. Regressors at point 2. plus $\cos(\lambda_i t)$, $\sin(\lambda_i t)$, with $\lambda_i = 2\pi i/24$ and $i = 1, \ldots, 6$.

Notice that the sinusoids at point 1. take care of the within-year seasonality, while those at point 3. model the within-day seasonality. These latter sinusoids are also supplemented with 24h-lagged prices (see below) that also help modelling the within-day seasonality.

Both vector autoregressive models and error correction mechanisms are explored. In particular, we regress:

Level 1-step: $\mathbf{y}_t$ on $\mathbf{y}_{t-1}, \mathbf{y}_{t-24}, \mathbf{y}_{t-168}, \mathbf{z}_t$;
Diff 1-step: $\Delta\mathbf{y}_t$ on $\mathbf{y}_{t-1}, \Delta\mathbf{y}_{t-1}, \Delta\mathbf{y}_{t-24}, \Delta\mathbf{y}_{t-168}, \Delta\mathbf{z}_t$;
Level 24-step: $\mathbf{y}_t$ on $\mathbf{y}_{t-24}, \mathbf{y}_{t-168}, \mathbf{z}_t$;
Diff 24-step: $\Delta_{24}\mathbf{y}_t$ on $\mathbf{y}_{t-24}, \Delta_{24}\mathbf{y}_{t-24}, \Delta_{24}\mathbf{y}_{t-168}, \Delta_{24}\mathbf{z}_t$.

The chosen rank and the actual root MSE (RMSE) are summarized in Table 2.

Three features appear evident from these figures: (i) the optimal rank of both PC and RRR models is very close to the full rank (51), indicating that almost all the information that the time series carry is relevant for forecasting; (ii) there is no clear indication about the choice of the algorithm, as the best algorithm for one-hour-ahead predictions is RRR while that for one-day-ahead predictions is PC; (iii) a large number of deterministic regressors is better than a small one.

Of course these regressors could have been supplemented with variables such as (lagged) oil prices, weather forecasts, and holidays dummies, that would certainly improve the in- and out-of-sample fit, but the main objective of this paper is proposing feasible techniques for forecasting this type of functional time series and testing them on real electricity auction data. The main features of the data are well captured by these regressors and lagged supply functions, and at this stage the fine-tuning of the models is not necessarily interesting.

As already mentioned, the above model selection was based on the out-of-sample RMSE of quantity increments, but since the mean absolute percentage error (MAPE) of the predicted function is easier to interpret and probably more eloquent the following discussion will be based on the latter loss measure.

Figure 3 depicts the out-of-sample MAPE as a function of time (first panel) and of price (second panel). It appears clear that the precision of the predictions vary significantly over time, but only slightly over price. In particular, the first half

Table 2 Out-of-sample root mean square error for the two algorithms and 12 models

	One-hour-ahead						One-day-ahead					
	Reg. 1.		Reg. 2.		Reg. 3.		Reg. 1.		Reg. 2.		Reg. 3.	
	Rank	RMSE	Rank	RMSE	Rank	RMSE	Rank	RMSE	Rank	RMSE	Rank	RMSE
RRR-Level	50	76.5	50	76.5	50	**67.1**	44	216.1	44	216.1	44	215.5
RRR-Diff	51	*73.7*	51	*73.7*	51	73.7	51	198.7	51	198.7	51	198.7
PC-level	50	80.9	50	80.9	50	72.7	41	*191.1*	41	*191.1*	41	**188.8**
PC-Diff	37	*73.7*	47	*73.7*	37	73.7	51	198.8	51	198.8	51	198.8

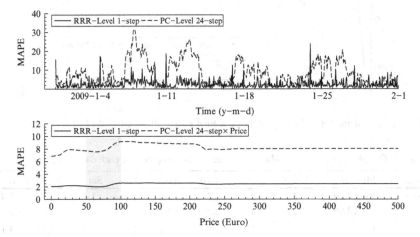

Fig. 3 Mean absolute percentage error of prediction as functions of time and price

of Jan-2009 seems to be harder to predict than the following part of that month. Indeed, those days are characterised by holidays and school vacations that were not explicitly modelled. As for the precision of the prediction at different points of the supply function, the quantities corresponding to the price interval [100, 200] are slightly more difficult to predict. Most observed SMPs are in the range [50, 100], and so this interval is the most interesting to predict. The MAPE in that interval is not particularly large: it is around 2% for one-hour-ahead predictions and some 8% for one-day-ahead predictions.

Figure 4 depicts the one-hour- and one-day-ahead functional prediction for an arbitrary auction chosen in the out-of-sample period, just to give a visual idea of the outcomes of the proposed algorithms.

Table 3 reports the MAPE computed for each day of the week. It reveals that supply functions are easiest to predict on Sundays and hardest to forecast on Mondays. The same table shows also that for one-hour-ahead predictions the RRR model on levels tend to be the best choice, while for one-day-ahead forecasts one should change the model according to the day of interest.

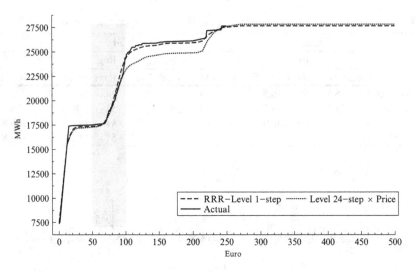

Fig. 4 Predicted and actual supply functions of Enel's competitors on Wed 14.01.2009 at 10am

Table 3 Out-of-sample mean absolute percentage error computed for each day of the week

Pred.	Method	Mon	Tue	Wed	Thu	Fri	Sat	Su	Ave.
1-step	RRR-Level	**2.8**	2.1	**2.5**	2.5	**2.3**	**2.4**	2.1	**2.4**
	RRR-Diff	2.9	**2.1**	2.8	2.3	2.3	2.8	2.0	2.5
	PC-Level	4.5	3.7	2.9	3.1	2.6	3.0	2.2	3.1
	PC-Diff	2.9	**2.1**	2.8	**2.3**	2.3	2.8	**2.0**	2.5
24-step	RRR-Level	11.3	9.6	**7.2**	**7.3**	6.6	8.2	9.1	**8.5**
	RRR-Diff	15.1	8.7	9.7	8.7	**4.1**	13.9	15.1	10.8
	PC-Level	**11.1**	11.6	7.2	11.6	9.3	**7.8**	**6.2**	9.2
	PC-Diff	15.1	**8.7**	9.7	8.7	4.2	13.9	15.1	10.8

6 Conclusions

We have introduced two different approaches to forecasting supply functions in electricity auctions. Accurate approximations of actual competitors' supply functions are indeed needed by all the generation companies bidding in the hourly (or semi-hourly) uniform-price auctions that characterise most electricity markets around the world. The two techniques are easy to implement and assure that the predictions share the same characteristics as the actual supply functions (i.e. positivity and non-decreasing monotonicity).

The application of the two techniques to the aggregate supply functions bid in the Italian day-ahead-market by Enel's competitors reveals that the predictions turn out to be accurate, but probably the optimal strategy should be adjusted to take into account the uncertainty about future supply functions. The proposed prediction algorithms, possibly supplemented with other relevant regressors, seem to represent a valuable tool for helping generation companies to design their bidding strategies in a more profitable way.

References

1. Bosco, B., Parisio, L., Pelagatti, M.: Strategic bidding in vertically integrated power markets with an application to the Italian electricity auctions. Energy Econ. **34**(6), 2046–2057 (2012). DOI 10.1016/j.eneco.2011.11.005
2. Bosq, D.: Linear Processes in Function Spaces: Theory and Applications. Springer, Berlin (2000)
3. Bunn, D., Farmer, E. (eds.): Comparative Models for Electrical Load Forecasting. Wiley, New York (1985)
4. Harvey, A., Koopman, S.J.: Forecasting hourly electricity demand using time-varying splines. J. Am. Stat. Assoc. **88**(424), 1228–1236 (1993)
5. Hortaçsu, A., Puller, S.L.: Understanding strategic bidding in multi-unit auctions: a case study of the Texas electricity spot market. RAND J. Econ. **39**(1), 86–114 (2008)
6. Hyndman, R., Shang, H.: Forecasting functional time series. J. Korean Stat. Soc. **38**, 199–211 (2009)
7. Hyndman, R., Ullah, M.: Robust forecasting of mortality and fertility rates: a functional data approach. Comput. Stat. Data Anal. **51**, 4942–4956 (2007)
8. Kastl, J.: Discrete bids and empirical inference in divisible good auctions. Rev. Econ. Stud. **78**(3), 974–1014 (2011)
9. Ramanathan, R., Engle, R.F., Granger, C., Vahid-Arahi, F., Brace, C.: Short-run forecasts of electricity loads and peaks. Int. J. Forecast. **13**, 161–174 (1997)
10. Ramsay, J.: When the data are functions. Psychometrika **47**, 379–396 (1982)
11. Ramsay, J., Silverman, B.: Functional Data Analysis, 2nd edn. Springer, Berlin (2005)
12. Reinsel, G., Velu, R.: Multivariate Reduced Rank Regression. Springer, Berlin (1998)
13. Wolak, F.: Identification and estimation of cost functions using observed bid data: an application to electricity markets, pp. 133–169. Cambridge University Press, Cambridge (2003)

A Hierarchical Bayesian Model for RNA-Seq Data

Davide Risso, Gabriele Sales, Chiara Romualdi, and Monica Chiogna

Abstract In the last few years, RNA-Seq has become a popular choice for high-throughput studies of gene expression, revealing its potential to overcome microarrays and become the new standard for transcriptional profiling. At a gene-level, RNA-Seq yields counts rather than continuous measures of expression, leading to the need for novel methods to deal with count data in high-dimensional problems.

We present a hierarchical Bayesian approach to the modeling of RNA-Seq data. The model accounts for the difference in the total number of counts in the different samples (sequencing depth), as well as for overdispersion, with no need to transform the data prior to the analysis. Using an MCMC algorithm, we identify differentially expressed genes, showing promising results both on simulated and on real data, compared to those of *edgeR* and *DESeq* (state-of-the-art algorithms for RNA-Seq data analysis).

1 Introduction

Next generation sequencing technologies have become widely used for measuring genome-wide transcription level, in what is called RNA-Seq. This technology produces short sequences, called "reads," consisting in a small number of DNA

D. Risso (✉)
Department of Statistics, University of California, Berkeley, CA
e-mail: davide.risso@berkeley.edu

G. Sales · M. Chiogna
Department of Statistical Sciences, University of Padua, Padua, Italy
e-mail: gabriele.sales@unipd.it; monica.chiogna@unipd.it

C. Romualdi
Department of Biology, University of Padua, Padua, Italy
e-mail: chiara.romualdi@unipd.it

M. Grigoletto et al. (eds.), *Complex Models and Computational Methods in Statistics*,
Contributions to Statistics, DOI 10.1007/978-88-470-2871-5_17,
© Springer-Verlag Italia 2013

bases (usually 25–100), that are mapped back to a reference genome. Subsequently, one counts the number of reads that fall within some pre-defined regions of interest (e.g., genes), leading to a discrete measure of expression (i.e., counts [27]).

For this reason, the Poisson model has been proposed to find differentially expressed (DE) genes, as it is the simplest model to deal with count data [5, 14, 26]. However, several authors have observed overdispersion when the experiment involves biological replicates (i.e., different individuals for each condition [1, 19]).

Robinson et al. [19] and Anders and Huber [1] propose a negative binomial (NB) distribution to model RNA-Seq data in a frequentist setting and they implement the approach in the Bioconductor packages *edgeR* and *DESeq*, respectively. Although the NB is a flexible distribution that allows the modeling of overdispersion, it is known that for small sample sizes the usual maximum likelihood estimator tends to underestimate the dispersion parameter [21]. For this reason, Robinson et al. [19] use a weighted conditional log-likelihood approach to shrink the estimates of the dispersion parameter towards a common value, mimicking an empirical Bayes solution [20]. They also propose to pull together the expression of all the genes and estimate a common dispersion parameter when the sample size is too small [21].

In a similar way, Anders and Huber [1] assume that the per-gene raw variance is a smooth function of the mean and pool the data from genes with similar expression strength to estimate the variance, using a local regression approach.

By virtue of the assay, the counts are not directly comparable between samples, since the total number of reads produced by the sequencer (known as *sequencing depth*) can vary between different runs. To deal with the difference in sequencing depth, both [1, 19] incorporate in the model a "size factor," which can be viewed as an offset in the model, that accounts for the difference in the library sizes. Finally, they consider an exact test for the comparison between two groups, as well as a likelihood ratio test for more general designs.

Hardcastle et al. [7] present an empirical Bayesian approach to differential expression in two-class and multi-class comparisons. By assuming an NB distribution, they enumerate all the possible patterns of differential expression and retain the pattern which is more likely to generate the data. In a nonparametric setting, Tarazona et al. [25] propose a test for differential expression in two-class comparisons based on the combination of log ratios and absolute differences between the mean expression of the two classes.

Hierarchical Bayesian models are a good alternative in this setting, since they can take advantage of the structure of the problem, allowing to borrow strength from the ensemble of the expression values [8]. This class of models has been successfully applied to microarray data both from an empirical [9, 13, 24] and a full Bayesian perspective [4]. Moreover, the negative binomial distribution arises as a Gamma–Poisson mixture model [11]; hence, one can model overdispersion in a hierarchical Bayes framework, using either a Gamma or a log-normal distribution as a prior for the mean parameter.

The article is organized as follows: in Sect. 2 we formulate the problem within the framework of generalized linear models (GLM) and we specify our proposed model; in Sect. 3 we perform a simulation study to assess the behavior of our proposed model; in Sect. 4 we show the results on a real dataset; finally, after a discussion in Sect. 5, we provide conclusions and future directions in Sect. 6.

2 Modeling Strategy

Most gene expression studies are designed to identify genes that are DE between two (or more) conditions, i.e., whose expression is different between the conditions of interest. GLM are a natural class of models to deal with these problems. They consider the gene expression level as the response and an ANOVA type design matrix that allows multiple class comparisons. Note that the GLM approach is able to model other types of studies, in which continuous covariates are involved. Hence, our model is not limited to class comparisons.

For each gene j, $j = 1, \ldots, p$, we model the expression values as realizations of a Poisson distribution. The expected values are specified in the following way:

$$\log(E[Y_j | X, \alpha, \beta_j]) = X\beta_j + \alpha, \tag{1}$$

where Y_j is the gene expression vector for gene j, X is the design matrix, β_j is a vector of parameters capturing the differences in mean among the classes, and α is a n-size vector of parameters used to capture the global difference among the samples due to the sequencing depth.

For the sake of clarity, consider a two-class comparison. The β_j parameter in (1) has two components: the first component, namely $\beta_{0,j}$, can be interpreted as the log-mean expression of gene j for the first class, while the second component, namely $\beta_{1,j}$, can be interpreted as the log-difference between the mean expression of gene j in the two classes. In this case, the design matrix will have two columns: an intercept and an indicator of the class membership for each sample. Note that more complex designs (e.g., multi-class comparison, matched samples, continuous response, ...) can be easily modeled by considering the proper design matrix.

Denoting with i, $i = 1, \ldots, n$, the samples, we consider the following prior distributions for the parameters:

$$\alpha_i \sim N(\mu_\alpha, \sigma_\alpha);$$

$$\beta_{0,j} \sim N(\mu_{\beta_0}, \sigma_{\beta_0});$$

$$1/\sigma_{\beta_0} \sim Ga(\psi_0, \phi_0);$$

where $\mu_\alpha = \mu_{\beta_0} = 0, \sigma_\alpha = 1, \psi_0 = \phi_0 = 10^{-2}$. As for $\beta_{1,j}$, we consider a mixture distribution, which allows us to model a situation in which most of the genes are not DE (i.e., $\beta_{1,j} = 0$),

$$\beta_{1,j} = (1 - w_j) V_j + w_j Z_j,$$

where

$$w_j \sim Bi(1, \pi);$$
$$Z_j \sim N(\mu_{Z_j}, \sigma_Z);$$
$$V_j \equiv 0;$$
$$\mu_{Z_j} \sim N(0, \sigma_{0Z});$$

where $\pi = 0.5$ and $\sigma_Z = \sigma_{0Z} = 1$.

To compute the posterior distribution of the parameter, we consider a Gibbs sampling algorithm, implemented using the JAGS software [17].

2.1 Differential Expression

In order for the model to be useful to researchers, there is the need for a way to declare a gene DE. One can use, in an empirical Bayes setting, a frequentist test statistic using the posterior distribution of β to estimate the variance in a way similar to [24].

Depending on the biological problem and on the sensitivity of the technology, all genes could be considered DE to some extent. In fact, as the technology improves, there is more sensitivity to capture even very small differences in gene expression. For this reason, some authors have argued that using a threshold on the fold-change while controlling for statistical variability is to be preferred to methods based solely on statistical significance [15, 28].

Here, we declare the gene j as DE based on the posterior probability of $\beta_{1,j}$. Given a predefined threshold t, we say that j is up-regulated if

$$Pr(\beta_{1,j} > t | Y) \geq 0.9; \tag{2}$$

analogously, we say that j is down-regulated if

$$Pr(\beta_{1,j} < -t | Y) \geq 0.9. \tag{3}$$

The choice of t has clearly a strong impact on the results; however, often researchers are interested in the "most DE" genes, and they use the DE statistic to rank the genes and to select the first ones as a base for subsequent analyses such as pathway analysis or gene set enrichment.

For this reason, if one has in mind a reasonable proportion of genes to expect as DE, one can use an appropriate quantile of the distribution of the posterior means of the $\beta_{1,j}$'s to choose t. For instance, in the simulation study, we use the 95th percentile when considering a scenario in which the 5% of the genes are simulated as DE.

3 Simulation Study

We simulate the data from our proposed model varying the values of the parameters to mimic different scenarios. We evaluate our model by considering the true and false positive rates (TPR and FPR, respectively), defining a gene as DE according to the rules in (2) and (3), choosing t as the quantile corresponding to the true proportion of DE genes (e.g., when simulating $\pi = 0.05$ in a balanced setting we will choose t as the 0.975 percentile in both (2) and (3)). We consider $p = 1,000$ genes and $n = 10$ samples in a two-class comparison in which the first five samples belong to class "1" and the others to class "2." We simulate according to the model in (1), with the following specifications: $\mu_\alpha = 5$, $\sigma_\alpha = \{0.2, 1\}$, $\mu_{\beta_0} = 0$, $\sigma_{\beta_0} = 1$. As for β_1, we fix the values to consider that a subset of genes is DE between the two conditions, with a log-fold-change of 2, either balanced (i.e., 50% up- and 50% down-regulated), moderately imbalanced (i.e., 75% up- and 25% down-regulated), or strongly imbalanced (i.e., 100% up-regulated). We let the proportion of DE genes vary in size, by simulating the hyperparameter $\pi = \{0.05, 0.1, 0.3\}$. The combination of these choices leads to a total of 18 possible scenarios. For each scenario, we simulated $B = 100$ datasets.

We compare our approach to two widely used state-of-the-art algorithms for differential expression of RNA-Seq data, namely *edgeR* [19] and *DESeq* [1].

For both algorithms, we followed the recommended pipeline, consisting in: (1) estimation of the "size factor" (analogous to our α parameter, or to a between-lane normalization), (2) estimation of the dispersion parameter, (3) testing for differential expression via an exact test based on the NB. For details, see the package vignettes at http://bioconductor.org. We define the true/false positives by considering a threshold of 0.05 on the (unadjusted) p-values.

Table 1 reports the results of the simulations for our proposed model. In general, the performance of our model is satisfying. In particular, the small number of false positives (FPR) means that the model is able to control for the type I error. The power of the test (TPR) is not extremely high; this can be due to the small sample size (five samples per class) or to the nature of our DE statistic, and other strategies to call for DE genes can be explored to find an optimal solution.

As expected, as σ_α increases, the model finds it more difficult to fit the data. In fact, when α is more variable, the true expression is more likely to be confounded with the differences in sequencing depth. For instance, the scenario in which $\sigma_\alpha = 1$ and $\pi = 0.3$, corresponding to a large proportion of DE genes in a highly variable experiment in terms of sequencing depth, is the only case which leads to an FPR greater than 0.05 (last three rows of Table 1).

Surprisingly, the test has more power for the imbalanced scenarios with respect to the balanced ones. This is an artifact due to the fact that the posterior distribution of $\beta_{1,j}$ tends to be shifted to the right.

Table 2 reports the results for *edgeR*. In all scenarios the test is anti-conservative, failing to control for the type I error.

Table 1 True positive rate (TPR) and false positive rate (FPR) of our proposed DE test in the 18 scenarios described in the main text

σ_α	π	d	TPR	s.e.	FPR	s.e.
0.2	0.05	0.5	0.548	0.067	0.010	0.004
0.2	0.05	0.25	0.581	0.067	0.010	0.003
0.2	0.05	0	0.648	0.069	0.009	0.003
0.2	0.1	0.5	0.608	0.057	0.015	0.005
0.2	0.1	0.25	0.641	0.051	0.017	0.005
0.2	0.1	0	0.714	0.048	0.015	0.004
0.2	0.3	0.5	0.729	0.039	0.033	0.010
0.2	0.3	0.25	0.746	0.035	0.034	0.009
0.2	0.3	0	0.811	0.034	0.031	0.007
1	0.05	0.5	0.429	0.083	0.013	0.004
1	0.05	0.25	0.464	0.083	0.013	0.004
1	0.05	0	0.541	0.080	0.012	0.004
1	0.1	0.5	0.520	0.066	0.022	0.008
1	0.1	0.25	0.541	0.070	0.023	0.007
1	0.1	0	0.621	0.069	0.020	0.006
1	0.3	0.5	0.689	0.039	0.061	0.017
1	0.3	0.25	0.698	0.042	0.057	0.014
1	0.3	0	0.760	0.043	0.046	0.011

The d parameter is the proportion of down-regulated genes among the DE

Table 2 True positive rate (TPR) and false positive rate (FPR) of *edgeR* DE test in the 18 scenarios described in the main text

σ_α	π	d	TPR	s.e.	FPR	s.e.
0.2	0.05	0.5	0.842	0.055	0.106	0.011
0.2	0.05	0.25	0.839	0.053	0.107	0.011
0.2	0.05	0	0.819	0.052	0.109	0.011
0.2	0.1	0.5	0.851	0.036	0.107	0.012
0.2	0.1	0.25	0.839	0.035	0.109	0.011
0.2	0.1	0	0.794	0.042	0.116	0.011
0.2	0.3	0.5	0.873	0.022	0.110	0.013
0.2	0.3	0.25	0.808	0.025	0.141	0.013
0.2	0.3	0	0.642	0.031	0.219	0.018
1	0.05	0.5	0.840	0.055	0.105	0.011
1	0.05	0.25	0.841	0.053	0.106	0.011
1	0.05	0	0.820	0.052	0.107	0.011
1	0.1	0.5	0.849	0.036	0.106	0.012
1	0.1	0.25	0.836	0.035	0.108	0.011
1	0.1	0	0.792	0.042	0.116	0.011
1	0.3	0.5	0.872	0.023	0.110	0.013
1	0.3	0.25	0.807	0.027	0.139	0.014
1	0.3	0	0.636	0.033	0.221	0.021

The d parameter is the proportion of down-regulated genes among the DE

Table 3 True positive rate (TPR) and false positive rate (FPR) of *DESeq* DE test in the 18 scenarios described in the main text

σ_α	π	d	TPR	s.e.	FPR	s.e.
0.2	0.05	0.5	0.319	0.114	0.020	0.005
0.2	0.05	0.25	0.315	0.105	0.020	0.005
0.2	0.05	0	0.284	0.100	0.021	0.005
0.2	0.1	0.5	0.405	0.072	0.020	0.005
0.2	0.1	0.25	0.381	0.081	0.021	0.005
0.2	0.1	0	0.313	0.072	0.024	0.006
0.2	0.3	0.5	0.524	0.039	0.022	0.007
0.2	0.3	0.25	0.433	0.038	0.031	0.007
0.2	0.3	0	0.202	0.032	0.058	0.010
1	0.05	0.5	0.080	0.080	0.008	0.005
1	0.05	0.25	0.081	0.076	0.008	0.005
1	0.05	0	0.071	0.077	0.008	0.005
1	0.1	0.5	0.118	0.083	0.008	0.005
1	0.1	0.25	0.108	0.081	0.008	0.005
1	0.1	0	0.080	0.070	0.009	0.006
1	0.3	0.5	0.208	0.099	0.008	0.005
1	0.3	0.25	0.166	0.084	0.013	0.009
1	0.3	0	0.063	0.049	0.027	0.016

The d parameter is the proportion of down-regulated genes among the DE

This algorithm is not sensitive to the higher variability in sequencing depth, being FPR and TPR similar independently of σ_α. However, a strong imbalance between up- and down-regulated DE genes increases the FPR and decreases the TPR, especially when the number of DE genes is large (i.e., $\pi = 0.3$ and $d = 0$).

Table 3 shows the results for *DESeq*. Strikingly, things are very different from *edgeR*: *DESeq* is able to control the type I error at the cost of a very low power.

In particular, *DESeq* is very sensitive to the high differences in sequencing depth and this leads to a TPR of less than 0.1 when $\sigma_\alpha = 1$ and $\pi = 0.05$.

Moreover, independently of the other parameters, when the DE genes are imbalanced the power decreases.

It is somewhat surprising that *edgeR* and *DESeq* lead to such different results. Indeed, they are both based on an NB model and they differ only by the way they estimate the size factor and the dispersion parameter.

Our feeling is that the difference depends on the way of estimating the dispersion: *edgeR* uses a likelihood-based approach [19], while *DESeq* exploits a local regression on the mean-variance scatterplot to estimate the dispersion [1]. It could be that this solution is more robust to the misspecification of the model (in terms of type I error) at the cost of less power.

To rule out the size factor estimation as the reason for the difference in the performance of *edgeR* and *DESeq*, we repeated the analyses on the simulated datasets after upper-quartile between-lane normalization (see [5]). For both models, the results are very similar to those of Tables 2 and 3 (data not shown), suggesting that the dispersion estimation is responsible for the difference in the performance of the two approaches.

4   Real Data

Bullard et al. [5] analyzed RNA-Seq data for two types of biological samples from the MicroArray Quality Control (MAQC) Project [23]: Ambion's human brain reference RNA (Brain), pooled from multiple donors and several brain regions, and Stratagene's universal human reference RNA (UHR), a mixture of total RNA extracted from ten different human cell lines. RNA-Seq was performed using Illumina's Genome Analyzer II high-throughput sequencing system. The data are summarized below: additional details about experimental design, pre-processing, and the associated qRT-PCR and microarray datasets can be found in [5].

We make use of the dataset "MAQC-2," where Brain and UHR RNA were sequenced each using a single library preparation and seven lanes distributed across two flow-cells (i.e., technical replicates). There are no biological replicates and library preparation effects are confounded with the extreme differential expression that one expects when comparing such different samples as Brain and UHR. Nonetheless, the availability of qRT-PCR measures for a subset of about 1,000 genes makes this a valuable benchmarking dataset.

Reads were mapped to the genome (GRCh37 assembly) using Bowtie [10], with unique mapping and up to two mismatches. Gene-level counts were obtained using the *union-intersection* (UI) gene model of [5]. Genes with an average read count below 10 were filtered out, retaining 12,340 out of 39,359 genes.

In the original MAQC paper [23], 997 genes were assayed by qRT-PCR, with four measures (i.e., technical replicates) for each of the Brain and UHR samples. This technology is known to yield accurate estimates of expression levels and it is used here as a gold standard for comparing the models. Following [5], we consider only the genes matching a unique UI gene, called present in at least three out of the four Brain and UHR runs, and having standard errors across the eight runs not exceeding 0.25. We found 638 genes in common with the RNA-Seq filtered genes and use this subset to compare expression measures between the technologies. The UHR/Brain expression log-fold-change of a gene is estimated by the log-ratio between the average of the four UHR measures and the average of the four Brain measures.

Table 4 shows the number of up- and down-regulated genes obtained after applying our proposed model to the real data. We found 764 up-regulated and 964 down-regulated genes (out of 12,340) in the Brain/UHR comparison. These numbers are much smaller than the DE genes identified by *edgeR* and *DESeq* and this could mean either lack of power of our test or the failure in the control of the type I error for the NB tests, or, more likely, a combination of these two effects.

We compute FPR and TPR, using the estimate of the UHR/Brain fold-change from qRT-PCR as the true value, considering only the subset of genes for which qRT-PCR data are available. Following the strategy described in [5], we consider three possible sets of genes: "non-DE," "DE," and "no-call," based on whether their qRT-PCR absolute log-fold-change is less than 0.2, greater than 2, or falls within the interval [0.2, 2], respectively. We obtain 208 "DE" and 81 "non-DE" genes.

Table 4 MAQC dataset: DE genes

	HB	edgeR	DESeq
Up	764	4,027	4,480
Down	964	4,011	4,501

Number of up- and down-regulated genes for our proposed model (HB), *edgeR* and *DESeq*

Table 5 MAQC dataset: true and false positive rates for the UHR/Brain comparison

	HB	edgeR	DESeq
TPR	0.856	1	0.995
FPR	0.000	0.741	0.716

True positive rates (TPR) and false positive rates (FPR) for the UHR/Brain comparison using our proposed model (HB), *edgeR* and *DESeq*

Table 5 shows that, at least for the subset of genes assayed by qRT-PCR, our model has less power but is able to control for the type I error, while the other two approaches lead to anti-conservative tests.

The results of our model and of *edgeR* in the real data are similar to those obtained in the simulations. Surprisingly, *DESeq* behaves very differently in the real data with respect to the simulations.

This is probably due to the library size estimation step that, as shown in Fig. 1, is biased for both *edgeR* and *DESeq*. In fact, for the MAQC dataset, the difference in sequencing depth is confounded with the biological difference between Brain and UHR, in which many more genes are expressed, being a pool of different tissues. Both *edgeR* and *DESeq* under-estimate the sequencing depth for the Brain samples and over-estimate it for UHR. Our model, simultaneously estimating α and β, is able to account for the limited differences in sequencing depth, while capturing the biological difference of interest between the two samples.

5 Discussion

A good alternative to frequentist approaches in the modeling of RNA-Seq is represented by hierarchical Bayesian models. Taking advantage of the structure of the problem, they borrow strength from the ensemble of the expression values and more effectively estimate the dispersion parameter. The log-normal prior allows to model experiments in which there is substantial overdispersion and outliers in the form of extremely high counts, having heavier tails than the Gamma distribution [3].

In the last few years, several authors have proposed methods for the identification of differentially expressed genes in RNA-Seq data. An exhaustive comparison of all the methods is beyond the scope of this article. However, we decided to compare our model to two popular approaches, namely *edgeR* and *DESeq*, as they are, to the best of our knowledge, the most widely used.

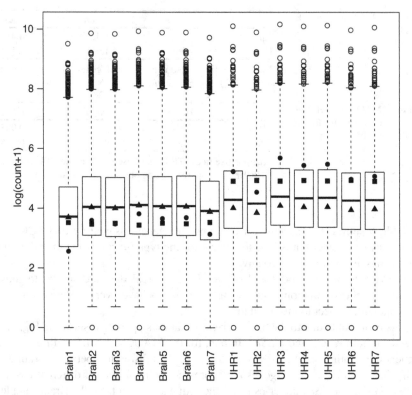

Fig. 1 *MAQC dataset: sequencing depth estimation.* Gene-level counts per lane for the MAQC dataset. The points represent the library sizes (scaled to scatter around the median) as estimated by: *square: edgeR* ; *circle: DESeq* ; *triangle*: our proposed model (α parameter)

Overall, our proposed model outperforms both *edgeR* and *DESeq* in the simulations. In fact, the high TPR of *edgeR* comes at the cost of the high FPR, meaning that the test fails to control for the type I error (recall that we used a threshold of 0.05 in the *p*-values distribution). On the other hand, *DESeq* controls for the type I error but has a very low power. Note that we are simulating from our proposed model, namely a Poisson-log-normal model, while both *edgeR* and *DESeq* are based on an NB. Therefore, we expect that our model will outperform these approaches in this setting, since *edgeR* and *DESeq* assume the overdispersion to be quadratic and we are simulating data with an exponential variance function [6, 11]. However, this simulation study was meant to be a confirmation of the goodness of the model when all the assumptions hold.

Moreover, in the simulation study we declare a gene DE according to equations (2) and (3), choosing *t* as the appropriate quantile, knowing the proportion of simulated DE genes. In real applications, this is clearly not possible and one has to choose a reasonable value for it, e.g., by fixing an appropriate threshold on the log-fold-change or by assuming a priori an expected proportion of DE genes.

Finally, one has to be careful in interpreting the results obtained in the real dataset, for two reasons: (1) the MAQC dataset is rather artificial, since it involves commercially available RNA and only technical replicates (with no biological variability); (2) the way we defined true and false positives is based on qRT-PCR and even if this is widely considered a gold standard for gene expression, one has to expect noise and measurement errors also in this technology. In other words, if RNA-Seq is more sensitive than qRT-PCR in the detection of DE genes, some of the genes that we declared "false positives" could be true positives failed to be detected by the qRT-PCR.

One important feature of the proposed model is that it can simultaneously estimate both biological effects (i.e., differential expression) and technical biases (e.g., difference in sequencing depth). The α parameter captures the difference in sequencing depth with no need of an a priori step of library size estimation, which is needed in both *edgeR* and *DESeq*. This step is essentially equivalent to a between-lane normalization and can be tricky, as discussed in [1, 5]. In particular using the sum of the counts to estimate the library size, as in Reads Per Kilobase of exon model per Million mapped reads (RPKM) of [16], can strongly bias differential expression [5].

Recently, Lee et al. [12] proposed a hierarchical Bayesian model for RNA-Seq. Their model works with "position-level" data, meaning that they model the read counts at each genomic position instead of considering gene-level summaries.

In this work, we consider gene-level data mainly for two reasons: (1) ease of interpretation and (2) computational convenience. In particular, considering position-level rather than gene-level data will add three orders of magnitude to the number of variables considered by the model. In fact, the human genome contains around 23,000 protein coding genes, with an approximate mean length of 2,000 DNA bases.

Here we refer to "genes" for the sake of clarity, but it is worth noting that one could use other "regions of interest" as count units. By considering, for instance, exon-level summaries, we are able in principle to identify differential usage of exons, evidence of differential isoform expression (see the discussion in [2]).

6 Conclusions

We have presented a hierarchical Bayesian GLM to model RNA-Seq data. Through a Gibbs sampler, we estimated the posterior distribution of the parameters and used them for inference on differential expression.

The model has the advantage of being very general and suitable for a wide range of studies, e.g., multi-class comparisons and designs with continuous covariate of interest.

It shows promising results when compared to existing approaches, both in simulations and in real data when using qRT-PCR as a gold standard.

The Gibbs sampling algorithm is computationally expensive, and this is a major concern in high-throughput studies, when several thousands of genes are analyzed at once. We will therefore take into consideration the possibility of exploiting the potential offered by the integrated nested Laplace approximations approach to overcome this limitation [22]. Moreover, future effort could be made on formal specification of the posterior probability of a differential expression statistic, in an empirical Bayes setting.

All the statistical analyses and simulations have been performed with R [18]. The Gibbs sampler was implemented in JAGS, version 3.1.0 [17], freely available at http://mcmc-jags.sourceforge.net/.

References

1. Anders, S., Huber, W.: Differential expression analysis for sequence count data. Genome Biol. **11**(10), R106 (2010)
2. Anders, S., Reyes, A., Huber, W.: Detecting differential usage of exons from RNA-seq data. Genome Res, online advanced access (2012)
3. Anscombe, F.: Sampling theory of the negative binomial and logarithmic series distributions. Biometrika **37**(3/4), 358–382 (1950)
4. Baldi, P., Long, A.: A Bayesian framework for the analysis of microarray expression data: regularized t-test and statistical inferences of gene changes. Bioinformatics **17**(6), 509–519 (2001)
5. Bullard, J., Purdom, E., Hansen, K., Dudoit, S.: Evaluation of statistical methods for normalization and differential expression in mRNA-Seq experiments. BMC Bioinformat. **11**(1), 94 (2010)
6. Bulmer, M.: On fitting the Poisson lognormal distribution to species-abundance data. Biometrics **30**(1), 101–110 (1974)
7. Hardcastle, T., Kelly, K.: baySeq: Empirical Bayesian methods for identifying differential expression in sequence count data. BMC Bioinformat. **11**(1), 422 (2010)
8. Ibrahim, J., Chen, M., Gray, R.: Bayesian models for gene expression with DNA microarray data. J. Am. Stat. Assoc. **97**(457), 88–99 (2002)
9. Kendziorski, C., Newton, M.A., Lan, H., Gould, M.N.: On parametric empirical Bayes methods for comparing multiple groups using replicated gene expression profiles. Stat. Med. **22**, 3899–3914 (2003)
10. Langmead, B., Trapnell, C., Pop, M., Salzberg, S.: Ultrafast and memory-efficient alignment of short DNA sequences to the human genome. Genome Biol. **10**(3), R25 (2009)
11. Lawless, J.: Negative binomial and mixed Poisson regression. Can. J. Stat. **15**(3), 209–225 (1987)
12. Lee, J., Ji, Y., Liang, S., Cai, G., Müller, P.: On differential gene expression using RNA-Seq data. Cancer Informat. **10**, 205 (2011)
13. Lönnstedt, I., Speed, T.: Replicated microarray data. Statistica Sinica **12**(1), 31–46 (2002)
14. Marioni, J., Mason, C., Mane, S., Stephens, M., Gilad, Y.: RNA-seq: an assessment of technical reproducibility and comparison with gene expression arrays. Genome Res. **18**(9), 1509 (2008)
15. McCarthy, D., Smyth, G.: Testing significance relative to a fold-change threshold is a TREAT. Bioinformatics **25**(6), 765 (2009)
16. Mortazavi, A., Williams, B., McCue, K., Schaeffer, L., Wold, B.: Mapping and quantifying mammalian transcriptomes by RNA-Seq. Nat. Methods **5**(7), 621–628 (2008)
17. Plummer, M.: JAGS: a program for analysis of Bayesian graphical models using Gibbs sampling. In: Proceedings of the 3rd International Workshop on Distributed Statistical Computing, pp. 20–22, March 2003

18. R Development Core Team: R: a language and environment for statistical computing. R Foundation for Statistical Computing, Vienna, Austria (2009). URL http://www.R-project.org

19. Robinson, M., McCarthy, D., Smyth, G.: edgeR: a Bioconductor package for differential expression analysis of digital gene expression data. Bioinformatics **26**(1), 139 (2010)

20. Robinson, M., Smyth, G.: Moderated statistical tests for assessing differences in tag abundance. Bioinformatics **23**(21), 2881 (2007)

21. Robinson, M., Smyth, G.: Small-sample estimation of negative binomial dispersion, with applications to SAGE data. Biostatistics **9**(2), 321 (2008)

22. Rue, H., Martino, S., Chopin, N.: Approximate Bayesian inference for latent Gaussian models by using integrated nested Laplace approximations. J. Roy. Stat. Soc. Ser. B (Methodolog.) **71**(2), 319–392 (2009)

23. Shi, L., Reid, L., Jones, W., et al.: The MicroArray Quality Control (MAQC) project shows inter- and intraplatform reproducibility of gene expression measurements. Nat. Biotechnol. **24**(9), 1151–1161 (2006)

24. Smyth, G.: Linear models and empirical Bayes methods for assessing differential expression in microarray experiments. Stat. Appl. Genet. Mol. Biol. **3**(1), 3 (2004)

25. Tarazona, S., García-Alcalde, F., Dopazo, J., Ferrer, A., Conesa, A.: Differential expression in RNA-seq: A matter of depth. Genome Res. **21**(12), 2213–2223 (2011)

26. Wang, L., Feng, Z., Wang, X., Wang, X., Zhang, X.: DEGseq: an R package for identifying differentially expressed genes from RNA-seq data. Bioinformatics **26**(1), 136 (2010)

27. Wang, Z., Gerstein, M., Snyder, M.: RNA-Seq: a revolutionary tool for transcriptomics. Nat. Rev. Genet. **10**(1), 57–63 (2009)

28. Wu, Z., Jenkins, B., Rynearson, T., Dyhrman, S., Saito, M., Mercier, M., Whitney, L.: Empirical Bayes analysis of sequencing-based transcriptional profiling without replicates. BMC Bioinformat. **11**, 564 (2010)